Modeling and Control of Uncertain
Nonlinear Systems with Fuzzy Equations
and *Z*-Number

Modeling and Control of Uncertain Nonlinear Systems with Fuzzy Equations and *Z*-Number

Wen Yu
CINVESTAV-IPN

Raheleh Jafari
University of Agder

IEEE PRESS

WILEY

For general information on our other products and services or for technical support, please contact our Customer Care Department within the United States at (800) 762-2974, outside the United States at (317) 572-3993 or fax (317) 572-4002.

Wiley also publishes its books in a variety of electronic formats. Some content that appears in print may not be available in electronic formats. For more information about Wiley products, visit our web site at www.wiley.com.

Library of Congress Cataloging-in-Publication Data is available.

Hardback: 9781119491552

Cover Design: Wiley
Cover Image: whiteMocca/Shutterstock

Printed in the United States of America.

V10011684_062719

To the daughters of Wen Yu: Huijia and Lisa
To the husband of Raheleh Jafari: Sina Razvarz

Modeling and Control of Uncertain
Nonlinear Systems with Fuzzy Equations
and *Z*-Number

Contents

List of Figures

List of Tables

Preface

Conventional mathematical tools, for example, difference equations, algebraic systems, and interpolation polynomials, have been used extensively for system modeling and parameter identification. Uncertain nonlinear system modeling is a mature subject with a variety of powerful methods and a long history of successful industrial applications [97][98]. The fuzzy modeling method is a popular tool for uncertain nonlinear system modeling. The fuzzy model usually comes from several fuzzy rules [182]. These fuzzy rules represent the controlled nonlinear system. Since any nonlinear system can be approximated by several piecewise linear systems (Takagi-Sugeno fuzzy model) or known nonlinear systems (Mamdani fuzzy model) [122], fuzzy models can approximate a large class of nonlinear systems while keeping the simplicity of the linear models. In this book, the authors discuss another type of fuzzy model. The basic idea is that many nonlinear systems can be expressed by linear-in-parameter models, such as Lagrangian mechanical systems [176]. The parameters of these models are uncertain and the uncertainties satisfy the fuzzy set theory [196]. In this way, the inconvenience problems in nonlinear modeling, such as complexity and uncertainty, are solved by the fuzzy logic theory and linear-in-parameter structure.

In recent years, many methods involving uncertainties have used fuzzy numbers [88][89][90][92][95][104][164][165][166][167], where the uncertainties of the system are represented by the fuzzy coefficients. The models are fuzzy equations or fuzzy differential equations (FDEs). The modeling process with the fuzzy equation is to find the fuzzy coefficients of the linear-in-parameter model such that the fuzzy equation can represent the uncertain nonlinear system. The application of the fuzzy equations and the FDEs are in direct connection with the nonlinear modeling and control.

Fuzzy control can be divided into direct and indirect methods [69]. Direct fuzzy control uses a fuzzy system as a controller, while indirect fuzzy control uses a fuzzy model to approximate the nonlinear system. The indirect fuzzy controller utilizes the simple topological structure and universal approximation ability of the fuzzy model. It has been widely used in uncertain nonlinear system control. The authors use indirect fuzzy control in this book.

The decisions are carried out on the basis of knowledge. In order to make the decision fruitful, the knowledge acquired must be credible. Z-numbers are associated with the reliability of knowledge [198]. Many fields related to the analysis of the decisions actually use the ideas of Z-numbers. Z-numbers are much less complex for calculation in comparison with nonlinear system modeling methods. The Z-number is an abundantly adequate number compared with the fuzzy number. Although Z-numbers are implemented in much of the literature, from the theoretical point of view this approach is not certified completely. In this book, the uncertainties are in the sense of Z-numbers.

From 2012 to 2016, the authors started to study the modeling and control of uncertain nonlinear systems with fuzzy equations. After four years of work, the results on the control of uncertain nonlinear systems with dual fuzzy equations, nonlinear system modeling with Bernstein neural networks, fuzzy modeling for uncertain nonlinear systems with fuzzy equations, numerical solution of fuzzy equations with Z-numbers, and numerical solution of fuzzy differential equations with Z-numbers were developed. These results have been published in a variety of journals and conferences. The authors wished to put together all these results within this book.

This book is organized as a textbook for a course on control and modeling of uncertain nonlinear systems. It could be used for self-learning. The level of competence expected for the reader is covered in the courses on nonlinear systems analysis, neural networks, and uncertain numbers.

Many people have contributed to shape the substance of the book. The first author, Wen Yu, would like to thank the financial supports of CONACyT-Mexico with the project 167428, and thanks his wife, Xiaoou Li, for her time and dedication. Without her this book would not have been possible. The second author, Raheleh Jafari, would like to express her sincere gratitude to her advisor Professor Wen Yu for his continuous support of her PhD study and research, and for his patience, motivation, enthusiasm, and immense knowledge. His guidance helped her throughout her research and writing of this book. Also, she would like to thank Professor Cristóbal Vargas Jarillo, Professor Alejandro Justo Malo Tamayo, Professor Sergio Salazar and Professor Tovar Rodriguez Julio Cesar. Last, but not least, the author thanks her husband, Sina Razvarz, for his time and dedication. Without him, this book would not have been possible.

Mexico   *Wen Yu and Raheleh Jafari*

1

Fuzzy Equations

1.1 Introduction

Fuzzy equations are a widespread problem in many applied fields, such as production planning, optimization decision, and artificial intelligence, in which establishing general and operable solving methods has received a remarkable amount of attention in academic circles. The fuzzy equation can be regarded as a generalized form of the fuzzy polynomial. Compared with normal fuzzy systems, fuzzy equations are easier to apply because the uncertainties are direct fuzzy parameters of the fuzzy equations.

1.2 Fuzzy Equations

The fuzzy method is a highly favorable tool for uncertain nonlinear system modeling. Fuzzy models approximate uncertain nonlinear systems with several linear piecewise systems (Takagi-Sugeno method) [182]. Mamdani models use fuzzy rules to achieve a good level of approximation of uncertainties [122]. When the parameter of an equation is changeable in the manner of a fuzzy set, this equation becomes a fuzzy equation [51]. When the parameters or the states of the differential equations are uncertain, they can be modeled with fuzzy differential equations (FDEs).

In comparison with normal systems, fuzzy equations are considered to be very non-complex. It is feasible for these equations to apply directly for nonlinear control. The approach of fuzzy control is associated with the design of appropriate nonlinear functions in the fuzzy equation. Fuzzy control with fuzzy equations requires the solution of the fuzzy equation. Several approaches are incorporated. In [70], the parametric form of fuzzy numbers is utilized and the original fuzzy equations using crisp linear systems are restored. In [52], the extension principle is implemented, which suggests that the coefficients can be either real or complex fuzzy numbers. However, the validation of the solution is not assured. In [8], the Newton technique is proposed. In [16], the solution

Modeling and Control of Uncertain Nonlinear Systems with Fuzzy Equations and Z-Number,
First Edition. Wen Yu and Raheleh Jafari.

of fuzzy equations is extracted using a fixed point methodology. The numerical solutions associated with fuzzy equations can be extracted using the iterative technique [108], interpolation technique [192], and Runge–Kutta technique [154]. They can also be implemented on partial differential equations (PDEs) and FDEs. In [185] the methodology of the Euler numerical technique is used to resolve the FDEs. The extension of classical fuzzy set theory in [83] results in obtaining the numerical solutions to FDEs. Some numerical approaches, such as the Nyström method [113] and the Runge–Kutta method [151] can also be implemented for resolving FDEs. The Laplace transform was utilized for second order FDEs in [20]. Several researchers have implemented the finite element technique, which has been used in the area of mechanics for solving a few specific PDEs [82]. Investigations by researchers revealing double non-traveling wave solutions associated with two systems of nonlinear PDEs are mentioned in [75]. The results of feedback control in reference to the wave equation are illustrated in [74], whereas open loop control concerning the wave equation is demonstrated in [115].

Both neural networks and fuzzy logic are considered to be the universal estimators that can estimate any nonlinear function to any notable precision [61][162][163]. Current outcomes demonstrate that the fusion methodology of these two different techniques appears to be highly efficient for nonlinear system identification [195]. Neural networks can also be implemented for resolving fuzzy equations [93][94]. A generalized fuzzy quadratic equation is resolved by utilizing neural networks, which is mentioned in [47]. In [103], the outcomes of [47] are elaborated into the fuzzy polynomial equation. Neural networks have been utilized in order to extract the solution of dual fuzzy equations, as illustrated in [95]. A matrix pattern associated with neural learning is proposed in [132]. However, these techniques are not general as they cannot resolve general fuzzy equations with neural networks. Also, they cannot generate fuzzy coefficients directly with neural networks [181]. In [67], a static neural network is proposed in order to resolve FDEs. In [15], the authors illustrate that the solutions to ordinary differential equations (ODEs) can be estimated with the help of neural networks. In [194], the neural approximations of ODEs to dynamic systems is implemented. In [118], dynamics neural networks are implemented for the approximation of the first order ODE. In [66], a feed forward neural network is suggested in order to resolve an elliptic PDE in 2D. The other methodology for solving a class of first order PDEs on the basis of multilayer neural networks is demonstrated in [79]. In [91], the neural network approach is used for solving strongly degenerate parabolic and Burgers–Fisher equations. The investigations of [129] laid down an unsupervised neural network to resolve the nonlinear Schrödinger equation. In [179], by employing a feed forward neural network, the controlled heat problem is solved.

1.3   Algebraic Fuzzy Equations

Some investigations have been carried out on algebraic fuzzy equations, but the existing methodologies calculate the roots of an algebraic fuzzy equation analytically. Also, there exists no analytical solution related to algebraic fuzzy equations having degree greater than 3 [48]. Thus utilization of numerical methodologies for such equations is essential. In [30], the author presented an algebraic fuzzy equation with degree n, including fuzzy coefficients and crisp variables, which is stated by

$$a_n x^n + \cdots + a_1 x + a_0 = 0 \tag{1.1}$$

where $0, a_0, a_1, \ldots, a_n \in E$ and $a_n \neq 0$. That is $P_n(x) = 0$, where $P_n(x) = \sum_{j=0}^{n} a_j x^j$ is mentioned as a polynomial of degree n. In order to determine the roots of the stated algebraic fuzzy equation an algorithm on the basis of Gauss–Newton technique produces a series that can converge under the condition that the modal value function includes a root or else the series can diverge. In addition, if the equation contains more than one root then the roots can be obtained via different initial vectors. In order to solve the algebraic fuzzy equation, the root and fuzzy zero are assumed to be unknowns, therefore using a series they can be determined. In the case that a fuzzy zero is provided, then only the root of the algebraic fuzzy equation should be extracted.

Consider the equations below

$$F(X) - B = 0, \quad F(X) = B \tag{1.2}$$

where B is an interval or fuzzy value and $F(X)$ is taken to be some interval or fuzzy function. These equations are not deemed to be equivalent. However, the major problem is linked to the conventional interval or fuzzy extension of the usual equation, which results in the interval or fuzzy equation $F(X) - B = 0$. An interval exists on the left hand side of this elongated equation, whereas a real valued zero exists on the right hand side. As it is not possible for the interval to be equal to the real value, this result is termed as "the interval equation right hand side problem". Minimal problems will occur while dealing with interval or fuzzy equations as $F(X) = B$, but in many issues its roots are termed as inverted intervals, i.e. in such a manner that $\bar{x} < \underline{x}$ [174].

The linear equation considered in [174] is mentioned as

$$ax = b \tag{1.3}$$

where its algebraically equivalent forms are denoted as

$$x = \frac{b}{a} \tag{1.4}$$

$$ax - b = 0 \tag{1.5}$$

where a and b are intervals. Suppose $[a] = (\underline{a}, \overline{a})$ and $[b] = (\underline{b}, \overline{b})$ are intervals. Therefore considering the case $[a] > 0$, $[b] > 0$, i.e. $\underline{a}, \overline{a} > 0$ and $\underline{b}, \overline{b} > 0$, the interval extension of Equation (1.3) is given by $(\underline{a}, \overline{a})(\underline{x}, \overline{x}) = (\underline{b}, \overline{b})$. This can be illustrated as $(\underline{ax}, \overline{ax}) = (\underline{b}, \overline{b})$. It is evident that the equivalence of the right as well as left hand sides of the equation is feasible only if $\underline{ax} = \underline{b}$ and $\overline{ax} = \overline{b}$, which are illustrated as

$$\underline{x} = \frac{\underline{b}}{\underline{a}}, \quad \overline{x} = \frac{\overline{b}}{\overline{a}}. \tag{1.6}$$

The interval extension of (1.4) can be regarded as

$$\underline{x} = \frac{\underline{b}}{\overline{a}}, \quad \overline{x} = \frac{\overline{b}}{\underline{a}}. \tag{1.7}$$

The suggested methodology can be utilized only in the uncomplicated cases of linear equations. Generally, choosing the suitable interval or fuzzy extensions in the nonlinear case is a complicated task.

Example 1.1   Assume $[a] = (3, 4)$ and $[b] = (1, 2)$. Hence from Equation (1.6) we can extract $\underline{x} = 0.333$, $\overline{x} = 0.5$, and from Equation (1.7) we can extract $\underline{x} = 0.25$, $\overline{x} = 0.666$. ∎

In [14], the decomposition methodology has been utilized for quadratic, cubic, and generalized higher order polynomial equations, and negative, or non-integral powers and random algebraic equations. The algebraic equations can be dealt with by using the decomposition methodology, which it supplies a crucial methodology in order to calculate the roots of polynomial equations, usually resulting in a very fast convergence. This methodology generally converges towards a precise solution.

In [33], a novel algorithm on the basis of the Adomian methodology is demonstrated in order to solve algebraic equations. This modernized algorithm computes the superior estimations related to the exact solution of algebraic equations, when compared with the standard Adomian methodology. A nonlinear equation is considered as follows

$$F(x) = 0 \tag{1.8}$$

which can be transformed to

$$x = F_0(x) + c_0 \tag{1.9}$$

where F_0 is taken to be a nonlinear function; c_0 is a constant. The Adomian methodology calculates x as a series

$$x = \sum_{i=0}^{\infty} x_i. \tag{1.10}$$

The decomposition of the nonlinear function is illustrated below

$$F(x) = \sum_{i=0}^{\infty} A_i \tag{1.11}$$

where A_i is considered to be an Adomian polynomial stated by

$$A_n(x_0, \ldots, x_n) = \left(\frac{1}{n!}\right)\left(\frac{d^n}{d\lambda^n}\right) F\left(\sum x_i \lambda^i\right)\bigg|_{\lambda=0}. \tag{1.12}$$

Substituting (1.10) and (1.11) into (1.9) results in

$$\sum_{i=0}^{\infty} x_i = \sum_{i=0}^{\infty} A_i + c_0 \tag{1.13}$$

where each term of the series $x = \sum_{i=0}^{\infty} x_i$, at par with Adomian method, can be computed using the relations

$$\begin{aligned} x_0 &= c_0 \\ x_1 &= A_0 \\ x_2 &= A_1 \\ &\quad\cdot \qquad \cdot \\ &\quad\cdot \\ &\quad\cdot \\ x_n &= A_{n-1} \end{aligned} \tag{1.14}$$

In calculating x by utilizing any software, since n increases, the number of terms in the expression for A_n increases and this results in the dissemination of round off errors. Also, the factor $\frac{1}{n!}$ mentioned in the formula related to A_n makes it minute, hence its contribution to x is not taken into account, and therefore the primary few terms related to the series $\sum_{i=0}^{\infty} x_i$ state the preciseness of the estimated solution. By taking this concept into consideration, [33] proposed a novel algorithm on the basis of the Adomian methodology in order to improve the preciseness significantly.

1.4 Numerical Methods for Solving Fuzzy Equations

1.4.1 Newton Method

In 1669, Isaac Newton introduced a novel algorithm [140] for solving a polynomial equation that was demonstrated on the basis of an example as $y^3 - 2y - 5 = 0$. To obtain a precise root of this equation, initially, a starting value should be assumed, where $y \approx 2$. By assuming $y = 2 + p$ and substituting it into the original equation, the following is obtained $p^3 + 6p^2 + 10p - 1 = 0$. As p is presumed to be minute, $p^3 + 6p^2$ is neglected in comparison with $10p - 1$. Also the previous equation generates $p \approx 0.1$, so a superior approximation of the root is $y \approx 2.1$. The repetition of this process is feasible and $p = 0.1 + q$ is extracted, and the substitution delivers $q^3 + 6.3q^2 + 11.23q + 0.061 = 0$, hence $q \approx -0.061/11.23 = -0.0054\ldots$, so a novel approximation of the root is $y \approx 2.0946$. The process needs to be repeated until the expected number of digits is achieved. In his methodology, Newton did not distinctly utilize the hypothesis of derivative, but only applied it to polynomial equations.

In [3], some effective numerical algorithms are demonstrated in order to solve the nonlinear equation $f(x) = 0$ on the basis of the Newton–Raphson methodology. The modified Adomian decomposition methodology is implemented for developing the numerical algorithms.

In [7], the Newton methodology is proposed in association with fuzzy nonlinear equations instead of standard analytical methodologies, as they are not appropriate throughout. The primary intention is to extract a solution for fuzzy nonlinear equation $F(x) = c$. Initially, the cited researchers mentioned the fuzzy nonlinear equation in parametric form as illustrated below

$$\begin{cases} \underline{F}(\underline{x}, \overline{x}, \alpha) = \underline{c}(\alpha) \\ \overline{F}(\underline{x}, \overline{x}, \alpha) = \overline{c}(\alpha) \end{cases} \tag{1.15}$$

so they resolved it by utilizing Newton's methodology.

In [77], iterative methods are illustrated to obtain a simple root δ, i.e. $f(\delta) = 0$ as well as $f'(\delta) \neq 0$, of a nonlinear equation $f(x) = 0$. The authors used the construction of some higher order modifications of Newton's method in order to resolve nonlinear equations. This construction maximizes the convergence order of prevailing iterative methodologies by one, two, or three units. This can be implemented in any iteration formula as well as per iteration. The resulting methodologies sum up only one additional function evaluation in order to maximize the order. This makes the computational effectiveness superior. This scheme can be employed for improving any prevailing iteration formula.

In [28], the investigators found the solution of

$$A_1 x \oplus A_2 x^2 \oplus \cdots \oplus A_n x^n = A_0 \tag{1.16}$$

where A_i and X^j belong to a fuzzy set (for $i = 1, \ldots n\, j = 0, 1, \ldots, n$). The fuzzy quantities are demonstrated in parametric form. The primary initiative is based on the conversion of the polynomial fuzzy coefficients into parametric form, thereby implementing Newton's technique on each limit. In the final phase, for finding the root, which is considered to be a fuzzy number, the α level sets of fuzzy coefficients on each limit are computed numerically.

The advantage of Newton's methodology is due to the convergence speed, once an adequately precise approximation is known. A drawback of this methodology is that a precise initial approximation of the solution is required to validate convergence. Another drawback of Newton's technique is due to the calculation and inversion of the Jacobian matrix $J(x)$ at each iteration.

The rapid convergence of Newton's technique is possible when an appropriate initial value is achieved. However, it is difficult to extract this kind of value, and this technique is also relatively expensive to implement [54]. Broyden's methodology is suggested to resolve this kind of equation. Broyden's methodology leans toward superlinear convergence. This methodology is selected since it is a superior alternative in comparison to Newton's technique, and it also minimizes the amount of computation at each iteration without markedly decreasing the speed of convergence. It substitutes the matrix A_{k-1}, whose inverse is directly evaluated at each iteration, in place of the Jacobian matrix J, and it minimizes the arithmetic operation $O(n^3)$ to $O(n^2)$ [54]. Instead of utilizing standard analytical methods, such as Buckley and Qu methods, which are not appropriate for resolving a system of fuzzy nonlinear equations taking into consideration that the coefficient is the fuzzy number, Broyden's technique is suggested for resolving fuzzy nonlinear equations. In [161], Broyden's technique is implemented in order to solve fuzzy nonlinear equations. Initially, fuzzy nonlinear equations are displayed in the parametric form, and they are resolved by utilizing the Broyden technique.

In [192], a new technique based on the Newton and Broyden methods is suggested for solving dual fuzzy nonlinear equations. The fuzzy nonlinear equations are transformed into parametric form and are then solved with Newton's method for initial iteration and Broyden's method for remainder of the iterations. The fuzzy coefficients are demonstrated in parametric form. This method reduces the calculation of the Jacobian matrix in every iteration.

Newton's method is relatively expensive, since the calculation of the Hessian on the first iteration is needed. Accordingly, the analytic expression for the second derivative is often complicated or intractable, requiring a lot of computation. The steepest descent method uses only first order information and does not deal with approximating second derivatives.

1.4.2 Steepest Descent Method

In [9], a numerical solution associated with fuzzy nonlinear equation $F(x) = 0$ is suggested using the steepest descent technique, where the fuzzy quantities are demonstrated in parametric form. The equation is represented by parametric form, as mentioned below

$$\begin{cases} \underline{F}(\underline{x}, \overline{x}, \alpha) = 0 \\ \overline{F}(\underline{x}, \overline{x}, \alpha) = 0 \end{cases}. \tag{1.17}$$

The function $K : R^2 \to R$ is stated as

$$K(\underline{x}, \overline{x}) = [\underline{F}(\underline{x}, \overline{x}, \alpha), \overline{F}(\underline{x}, \overline{x}, \alpha)]^2. \tag{1.18}$$

The technique of steepest descent characterizes a local minimum for two variable function K. The technique of steepest descent is stated as:

1) Find K at an initial approximation $X_0^\alpha = (\underline{x}_0^\alpha, \overline{x}_0^\alpha)$.
2) Determine a direction from $X_0^\alpha = (\underline{x}_0^\alpha, \overline{x}_0^\alpha)$ that causes a decrease in the value of K.
3) Shift a suitable amount in this direction and consider the new value $X_1^\alpha = (\underline{x}_1^\alpha, \overline{x}_1^\alpha)$.
4) Repeat sequence 1 via 3 with X_0^α replaced by X_1^α.

The steepest descent technique converges only linearly to the solution, but in general it converges even for weak initial approximations [53]. Even though the steepest descent technique does not need a superior initial value, its draw-back is due to its low convergence speed. Genetic algorithms have been found to provide a rapid convergence to a near optimum solution in many types of problems. The genetic algorithm method has better training performance than the steepest descent method.

1.4.3 Adomian Decomposition Method

The Adomian decomposition method was initially laid down by George Adomian in [13]. In [2], the standard Adomian decomposition is implemented on the simple iteration technique in order to resolve the equation $f(x) = 0$, where $f(x)$ is a nonlinear function, and the convergence related to the series solution is proved. Initially, the nonlinear equation is transformed into canonical form, then the Adomian technique computes the solution, which is at par with the series form. As practically all the terms associated with the series are not possible to determine, hence the estimation of the solution is obtained from the truncated series. Therefore, the convergence related to the truncated series is usually very rapid.

Babolian et al. [31] modified the standard Adomian technique mentioned in [2] in order to solve nonlinear equation $f(x) = 0$ to acquire a sequence of approximations related to the solution with approximate superlinear convergence. They employed Cherruault's definition [59] and took into consideration the order of convergence related to the technique [32].

In [149], a potential numerical algorithm in order to solve fuzzy polynomial equations $\sum_{i=1}^{n} a_i x^i = c$ on the basis of Newton's technique is demonstrated, where x and c are considered to be fuzzy numbers, and all coefficients are taken to be fuzzy numbers. The modified Adomian decomposition methodology is implemented for the construction of the numerical algorithm. Initially, the fuzzy polynomials are illustrated in parametric form and then resolved using the Adomian decomposition technique.

In [189], the Shanks transformation is employed on the Adomian decomposition technique in order to resolve nonlinear equations so as to improvise the preciseness of the approximate solutions. The numerical results demonstrate that the implementation of this technique in similar conditions generates more appropriate solutions to the nonlinear equations when compared with those extracted from the Adomian decomposition technique. The Shanks transform is an effective approach that can speed up the convergence rate of the series.

In [131], an effective extension of Newton's method to the fuzzy polynomial is proposed using a modified Adomian decomposition technique in the form of $\sum_{i=1}^{n} b_i x^i = d$, where x, d, and b are fuzzy numbers. The fuzzy polynomials are written in a parametric form and then are resolved by the Adomian decomposition method.

The advantage of the Adomian decomposition technique is that it can provide analytical approximations to solutions of nonlinear equations without supposing that the system has weak nonlinearities. The major drawback of the Adomian decomposition technique is the complex and difficult procedure needed to compute the Adomian polynomials. The ranking method is simple and inexpensive.

1.4.4 Ranking Method

The ranking methodology was primarily laid down by Delgado et al. [64]. In [170], the researcher obtained the real roots of the polynomial equation, which has been demonstrated as follows

$$C_1 x + C_2 x^2 + \cdots + C_n x^n = C_0 \tag{1.19}$$

where $x \in R$ as well as $C_0, C_1, \ldots, C_n$ are taken to be fuzzy numbers. In [170] the fuzzy polynomial equation is converted to a system of crisp polynomial equations. This conversion takes place using the ranking method on the basis of three parameters: walue, ambiguity and fuzziness. The obtained system of crisp polynomial equations is resolved numerically.

In [144], the conceptual content of a ranking method is suggested in order to extract the real roots associated with a dual fuzzy polynomial equation, which has been illustrated as follows

$$A_1 x \oplus A_2 x^2 \oplus A_n x^n = B_1 x \oplus B_2 x^2 \oplus B_n x^n + d \tag{1.20}$$

where $x \in R$ as well as $A_1, \ldots, A_n, B_1, \ldots, B_n, d$ are denoted as fuzzy numbers. The dual fuzzy polynomial equations are converted to a system of crisp dual polynomial equations. This conversion is carried out by utilizing ranking methodology on the basis of three parameters, namely value, ambiguity, and fuzziness.

In [145], the real roots corresponding to the polynomial equation, $A_1 x + A_2 x^2 + \cdots + A_n x^n = A_0$, is obtained by utilizing the ranking method considering fuzzy numbers, where $x \in R$ as well as $A_0, A_1, \ldots, A_n$ are denoted as fuzzy numbers. In the referred paper, the ranking methodology is utilized for real roots of the dual polynomial equations as mentioned below

$$A_1 x + A_2 x^2 + \cdots + A_n x^n = B_1 x + B_2 x^2 + \cdots + B_n x^n + d \tag{1.21}$$

where $x \in R, A_1, \ldots, A_n, B_1, \ldots, B_n$ as well as d are considered to be fuzzy numbers.

In [146], the ranking technique is implemented in order to obtain the real roots of an interval type 2 dual fuzzy polynomial equation $A_1 x + A_2 x^2 + \cdots + A_n x^n = B_1 x + B_2 x^2 + \cdots + B_n x^n + d$, where $x \in R$ and the coefficients $A_1, \ldots, A_n, B_1, \ldots, B_n$ as well as d are termed as interval type 2 fuzzy numbers. The type 2 dual fuzzy polynomial equation is converted into a system at par with the crisp type 2 dual fuzzy polynomial equation. The conversion is done by the ranking of the fuzzy numbers on the basis of three parameters, namely value, ambiguity, and fuzziness.

It has been revealed that solutions in correspond to three parameters value, ambiguity, and fuzziness are not sufficient to generate solutions. Hence in [147], a novel ranking methodology is suggested in order to eradicate the intrinsic weakness. The novel ranking methodology, which is incorporated with four parameters is then implemented in the interval type 2 fuzzy polynomials, covering the interval type 2 of fuzzy polynomial equations, dual fuzzy polynomial equations as well as the system of fuzzy polynomials. The effectiveness of the novel ranking methodology is numerically considered in the triangular fuzzy numbers as well as the trapezoidal fuzzy numbers.

The main disadvantage of the ranking method is that it can be applied only when membership functions are known. Approximation methods such as fuzzy neural networks are also effective tools for overcoming the limitations of the other numerical methods. The major advantage of using fuzzy neural networks is training the large amount of data sets, quick convergence, and high accuracy.

1.4.5 Intelligent Methods

1.4.5.1 Genetic Algorithm Method

A genetic algorithm for resolving the fuzzy equation $P(x) = y$ is demonstrated in [46], where x and y are considered to be k sampled real fuzzy numbers and P is taken to be a fuzzy function relying on x. The motivation is to obtain a suitable value of the fuzzy argument x in such a manner that the calculated value of the polynomial, $P(x)$, is very much adjacent to the supplied target value y. The presented genetic algorithm utilizes a distinct demonstration of the fuzzy numbers, which permits the implementation of simple genetic operators. The algorithm is self-sufficient for finding multiple solutions associated with the fuzzy equations. However, no method has been utilized for an identical problem involved in the area of neural networks that can be taken possession of. Because of the distinct discrete criteria of the fuzzy arithmetic, the single realistic approach for resolving this problem is to design a dedicated genetic algorithm [127].

A genetic algorithms methodology for resolving the linear and quadratic fuzzy equations $Ax = B$ as well as $Ax^2 + Bx = C$, where A, B, C and x are considered to be fuzzy numbers, is mentioned in [124]. The methodology based on the genetic algorithms primarily begins with a set of random fuzzy solutions. After that, in each generation of genetic algorithms, the solution candidates converge to the superior fuzzy solution. In the suggested methodology the final obtained solution is not only restricted to fuzzy triangular but it can also be a fuzzy number. In this methodology, in order to obtain the best fuzzy solution associated with a fuzzy equation, initially a solution is required to be converted to its chromosome demonstration. The solution candidates for the fuzzy equations are transformed to their level sets demonstration, which are computable by genetic algorithms.

In [120], genetic algorithms are implemented for solving fuzzy equations without stating membership functions related to fuzzy numbers. The extension principle, interval arithmetic, α cut operations, or a penalty technique were not used in order to deal with the problem. An important matter for using genetic algorithms in order to extract a better solution associated with the problem is the parameter settings, which include the probability of crossover, the probability of mutation, and the number of generations. The fuzzy concept related to the genetic algorithm scheme is different, but generates superior solutions in comparison with classical fuzzy techniques.

In [125], a genetic algorithm is used for resolving nonlinear equations of the form $g(x) = 0$, where x and $g(x)$ may be real, complex, or vector quantities. At first, $g(x) = 0$ is transformed into a minimization problem, then a genetic algorithm is applied for finding the minimum. The method is extended to systems of nonlinear equations.

The genetic algorithm represents the most consistent results in terms of accuracy and convergence but it is computationally very expensive. The modified Adomian provides acceptable results and converges rapidly to the numerical solution. The Adomian decomposition method is less expensive than the genetic algorithm method.

1.4.5.2 Neural Network Method

In [49] neural network was employed for solving the fuzzy linear equation

$$AX = C \tag{1.22}$$

where A, B, and X are considered to be triangular fuzzy numbers. Taking into account certain values of A and C, Equation (1.22) generates no solution for X [52]. The training of a neural network in order to solve Equation (1.22) was mentioned by the researchers in [49], considering that zero is not at par with the support of A. The investigation was carried out considering neural network solutions termed as Y and X^*. When there are no restrictions concerned with the weights of the network, then the neural network output will be Y. The non-existence of the relationship between Y and X was proved and validated by utilizing computer analysis. X^* is the solution of the neural network, taking into consideration that certain sign restrictions are set on the weights. X^* is illustrated to be an approximation, which is named a new solution of fuzzy equations. It has been displayed by using $X \leq X^*$.

The evolutionary algorithm, as well as a neural network in combination, has been utilized for solving the fuzzy equation, which was referred to in [50] as follows

$$AX \oplus B = C \tag{1.23}$$

where A, B, C and X are termed as triangular fuzzy numbers. The first solution X_c related to Equation (1.23) is stated to be the classical solution that utilizes α-cut and interval arithmetic for obtaining X_c.

Example 1.2 Assume $[A] = (1, 2, 3)$, $[B] = (-3, -2, -1)$ and $[C] = (3, 4, 5)$. Employing the intervals in the fuzzy equation generates

$$(1 + \alpha)\underline{X}_c^\alpha + (-3 + \alpha) = (3 + \alpha)$$
$$(3 - \alpha)\overline{X}_c^\alpha + (-1 - \alpha) = (5 - \alpha) \tag{1.24}$$

where $[X_c]^\alpha = (\underline{X}_c^\alpha, \overline{X}_c^\alpha)$. Then

$$\underline{X}_c^\alpha = \frac{6}{1 + \alpha}$$
$$\overline{X}_c^\alpha = \frac{6}{3 - \alpha} \tag{1.25}$$

can be extracted. However $[\underline{X}_c^\alpha, \overline{X}_c^\alpha]$ does not state a fuzzy number as $\underline{X}_c^\alpha(\overline{X}_c^\alpha)$ is a decreasing (increasing) function of α. Occasionally X_c prevails and sometimes will not exist. ∎

By the fuzzification of the crisp solution $(c - b)/a, a \neq 0$, the other solution is extracted. $(C - B)/A$ represents the fuzzified solution, taking into account that zero is not at par with the support of A. For the evaluation of the fuzzified solution two approaches have been suggested. The primary approach generates the solution X_e by utilizing the extension principle, and the secondary approach generates the solution X_I by the means of α-cut and interval arithmetic. X_e can be achieved as shown below

$$X_e = \min\{\Pi(a, b, c) | (c - b)/a = x\} \tag{1.26}$$

where $\Pi(a, b, c) = \min\{A(a), B(b), C(c)\}$. For obtaining α-cut of X_e the process is described as follows

$$\underline{X}_e^\alpha = \min \left\{ \left. \left| \frac{c-b}{a} \right| \right| a \in [A]^\alpha, b \in [B]^\alpha, c \in [C]^\alpha \right\}$$

$$\overline{X}_e^\alpha = \max \left\{ \left. \left| \frac{c-b}{a} \right| \right| a \in [A]^\alpha, b \in [B]^\alpha, c \in [C]^\alpha \right\} \tag{1.27}$$

where $[X_e]^\alpha = (\underline{X}_e^\alpha, \overline{X}_e^\alpha)$. The solution X_I can be calculated as follows

$$[X_I]^\alpha = ([C]^\alpha - [B]^\alpha)/[A]^\alpha. \tag{1.28}$$

The original fuzzy equation may or may not be solved by $X_e(X_I)$. Taking into account some fuzzy equations, X_e is mathematically too complex to extract, so in [50] an evolutionary algorithm was implemented for estimating their α-cuts. The method in the paper can be normalized for interacting with fuzzy problems, evolutionary algorithms, and neural networks. There are disadvantages in the method that were mentioned in [50]. The method is exclusively meant for symmetric fuzzy numbers, and it computes just the upper bound and the lower bound of the fuzzy numbers, avoiding the center part.

An architecture of the fuzzy neural network that is suggested in order to obtain a real root of the fuzzy polynomials is illustrated in the form [10]

$$A_1 x + \cdots + A_n x^n = A_0 \tag{1.29}$$

where $x \in R$ as well as $A_0, A_1, \ldots, A_n \in E$. A learning algorithm associated with the cost function in order to adjust the crisp weights has been suggested. The methodology mentioned in [10] has drawbacks. It was solely capable of extracting a crisp solution of fuzzy polynomials, and this neural network cannot extract a fuzzy solution.

In [103], the researchers obtained the approximate solution of the following fuzzy polynomial having degree n

$$A_1 x + \cdots + A_n x^n = A_0 \tag{1.30}$$

where $A_0, A_1, \ldots, A_n, x \in E$. They laid down two types of neural networks for approximating the solution of Equation (1.30), namely feedforward (static) as well as recurrent (dynamic) models. The corresponding algorithm of both neural networks is based on the least mean square. The difference between the two neural networks is that the dynamic neural network has superior robustness than the static neural network. The technique, illustrated in [103], is sufficient to find an approximate solution at par with the special case of a fuzzy equation, but not the generalized case.

The general fuzzy equation known as a dual fuzzy equation [192] was illustrated in [95]. Normal fuzzy equations have fuzzy numbers solely on one side of the equation. However, dual fuzzy equations have fuzzy numbers on both sides of the equation. Since it is not possible to move the fuzzy numbers from one side to the other [108], dual fuzzy equations are more generalized and complex. In [95], the existence of the solutions related to the dual fuzzy equations is analyzed, and is incorporated with the controllability problem of fuzzy control [58]. Afterward, two kinds of neural networks for the approximation of the solutions related to dual fuzzy equations were demonstrated, namely the static and dynamic models.

In [133], an architecture of fuzzy neural networks was suggested for solving dual fuzzy polynomial equations. A learning algorithm of fuzzy weights of two layers of feed-forward fuzzy neural networks is used whose input–output relations are defined by the extension principle.

In [102], a dynamic neural network is proposed for solving a dual fuzzy polynomial and is demonstrated as follows

$$a_1 x + \cdots + a_n x^n = b_1 x + \cdots + b_n x^n + d \qquad (1.31)$$

where $a_1, \ldots, a_n, b_1, \ldots, b_n$ and d belong to a fuzzy set. The neural network is trained by a back propagation type learning algorithm that has five layers where connection weights are crisp numbers. The important advantage of this methodology is that it can greatly reduce the size of calculations and generate high accuracy of the numerical solution.

Fuzzy linear regression analysis has become popular with investigators and is a standard model for analyzing data vagueness phenomena. It is utilized to generate a suitable linear relation between a dependent variable and various independent variables in a fuzzy environment.

1.4.5.3 Fuzzy Linear Regression Model

Generally, there exist two techniques in fuzzy regression analysis namely a linear programming based technique [155][172][183][184] and a fuzzy least squares technique [171][65]. The primary technique relies on diminishing fuzziness at par with optimal criteria. The secondary technique utilizes least square errors at par with fitting criteria. As illustrated in [191], the benefit of the primary technique is its simplicity of programming as well as calculation, whereas in the fuzzy least squares technique it is its minimal degree of fuzziness between the observed and approximated values. Currently, the least total square error of the spread values is utilized as the fitting criteria as well as an advanced mathematical programming methodology in which the predictability of the primary technique can be improved and the calculation complication of the secondary technique can be minimized [137].

Fuzzy linear regression was initially proposed by Tanaka et al. [184]. The main intention was to minimize the total spread of the fuzzy parameters relating to the support of the approximated values that enclose the support of the observed values at par with a certain α level. Even though this concept was later modified by Tanaka et al. [183], their model is deemed to be very responsive to data. It can generate infinite solutions as well as the spread of the approximated values, since it becomes wider as more data are piled up in the model.

In [172], a fuzzy linear regression model is generated in the form of $Y_i = A_0 + A_1 x_i$ at par with the fuzzy output as well as fuzzy parameters taking into consideration the mathematical programming problem by utilizing three indices concerned with the equalities between fuzzy numbers. Three patterns of multi-objective programming problems in order to extract fuzzy linear regression models are laid down related to the three indices. Linear programming relies on an interactive decision making method in order to extract the convenient solution at par with the decision making for formulating the multi-objective programming problem. The technique implied in [172] can generate an infinite number of solutions via repeated observations. Therefore, the mentioned technique is able to generate crisp coefficients. By repeated observations this technique results in redundant constraints. Hence, all observations cannot contribute to the computation of the model. In [172], fuzzy linear regression experiences crisp coefficients, redundant constraints, and the possibility of an infinite number of solutions. To deal with the possibility of an infinite number of solutions, the approximation point of the centers at par with the fuzzy coefficients can be computed using the available data, which is implemented into fuzzy linear regression algorithms. In [173], it was illustrated that the least squares technique can be utilized as a point approximation to center the fuzzy coefficients, and this is employed in the Tanaka technique. Two advantages are linked to the use of a point approximation to center the fuzzy coefficients. Primarily, if the researcher selects a point approximation that is distinctively defined, then the possibility of an infinite number of solutions is eliminated. Secondly, point approximations permit all data points to contribute information in the fuzzy linear algorithm. Hence, under repeated observations, the utilization of point approximations associated with the center of the fuzzy coefficients tackles some of the problems imparted by redundant constraints.

Nasrabadi et al. [138] utilized a multi-objective programming concept in order to illustrate the linear regression coefficients $Y_i = A_0 + A_1 X_i + \cdots + A_n X_{in}$, $i = 1, \ldots, m$ where $X_{ij} = (x_{ij}, r_{ij})$, $A = (a_j, \alpha_j)$ as well as $Y_i = (y_i, \beta_i)$ are considered to be symmetric fuzzy numbers. In the mentioned fuzzy regression work, powerful predictions are developed on the basis of the fuzzy number parameters.

In [134], the fuzzy linear regression $Y_i = A_0 \oplus A_1 x_{i1} \oplus A_2 x_{i2} \oplus \cdots \oplus A_n x_{in}$ as well as fuzzy polynomial regression $Y_i = A_{l0} \oplus \sum_{j=1}^{n} A_{lj} x_{ij} \oplus \sum_{j=1}^{n} \sum_{k=1}^{n} A_{ljk} x_{ij} x_{ik} \oplus \ldots$, where input units are taken to be crisp numbers and the output unit is taken to be a fuzzy number. The proposed technique relies on a neural network model in order to extract the estimation of the regression coefficients. A more generalized pattern of fuzzy polynomial regression was revealed in [148].

In [148], a polynomial fuzzy regression model at par with fuzzy independent variables, as well as fuzzy parameters based on the following form, is illustrated as

$$Y_i = A_{l0} \oplus \sum_{j=1}^{n} A_{lj}X_{ij} \oplus \sum_{j=1}^{n}\sum_{k=1}^{n} A_{ljk}X_{ij}X_{ik} \oplus \dots \tag{1.32}$$

where i denotes the different observations, $X_{i1}, X_{i2}, \dots, X_{in}$ are coefficients, and Y_i is considered to be a fuzzy number. A fuzzy neural network model is utilized to extract an estimate related to the fuzzy parameters in a statistical sense. This technique permits the development of nonlinear regression models, along with general fuzzy number inputs, outputs, and parameters. The suggested technique consists of numerous properties. Initially, it can use non-triangular fuzzy observations. Furthermore, the fuzzy neural network technique performs perfectly with respect to the sum of squared errors and accuracy of approximation.

In order to obtain a fuzzy polynomial interpolation having degree n, it is essential to extract n fuzzy coefficients. So as to find these coefficients, it is a requirement to solve $2n \times 2n$ equations, which is very complicated in terms of large values of n. Also sometimes it is barred of a fuzzy solution [71]. In [123], an innovative estimation algorithm for fuzzy polynomial interpolation by utilizing the artificial bee colony algorithm for interpolating fuzzy data is demonstrated. It is assumed that $X = \{x_1, \dots, x_m\}$ is a set of m distinct points associated with R and $F = \{y_1, \dots, y_n\}$ is the value of a triangular fuzzy function f at the point $x_i, i = 1, \dots, n$. The polynomial below of m degree is taken into account

$$p_m(x) = \sum_{j=0}^{m} a_j x^j = \sum_{j=0}^{m} (\underline{a_j}(r), \overline{a}_j(r)) x^j \tag{1.33}$$

where a_j is a trapezoidal fuzzy number at par with parametric form $(\underline{a_j}(r), \overline{a}_j(r))$ for $j = 0, 1, \dots, m$. The experimental data are considered to be $(x_1, y_1), (x_2, y_2), (x_n, y_n), x_i \in X$, and $y_i \in F$ (given that $n > m + 1$).

In order to compare the efficiency of the numerical methods to approximate the solution of dual fuzzy equations the examples below are demonstrated.

Example 1.3 A water tank system contains two inlet valves q_1 and q_2, as well as two outlet valves q_3 and q_4, see Figure 1.1. The areas of the valves are uncertain, $A_1 = G(0.021, 0.023, 0.024)$, $A_2 = G(0.008, 0.018, 0.038)$, $A_3 = G(0.012, 0.013, 0.015)$, and $A_4 = G(0.038, 0.058, 0.068)$. The velocities of the flow (controlled by the valves) are $g_1 = \left(\frac{\vartheta}{10}\right)e^{\vartheta}$, $g_2 = \vartheta cos(\Pi\vartheta)$, $g_3 = cos\left(\frac{\Pi\vartheta}{8}\right)$, and $g_4 = \frac{\vartheta}{2}$. If the outlet flow is aimed to be $q = (4.088, 6.336, 36.399)$, what is the quantity of the control variable ϑ?

The mass balance of the tank is [177],

$$\rho A_1 g_1 \oplus \rho A_2 g_2 = \rho A_3 g_3 \oplus \rho A_4 g_4 \oplus q \tag{1.34}$$

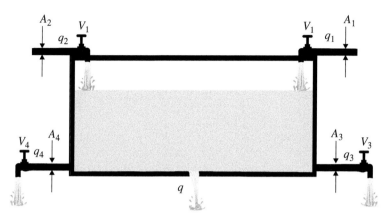

Figure 1.1 Water tank system.

Table 1.1 Approximation errors of the water tank.

k	Newton	Descent	Genetic	Decomposition	Ranking	NN
1	0.186	0.168	0.334	0.140	0.310	0.439
2	0.296	0.260	0.247	0.223	0.238	0.323
3	0.361	0.326	0.130	0.180	0.119	0.217
⋮	⋮	⋮	⋮	⋮	⋮	⋮
119	0.079	0.052	0.045	0.034	0.029	0.003
120	0.075	0.049	0.039	0.030	0.025	0.002

where ρ is the density of water. The exact solution is $\vartheta_0 = 2$ [177]. To approximate the solution, six popular methods are used: the Newton method, the steepest descent method, the genetic algorithm method, the Adomian decomposition method, the ranking method, and the neural network method. The errors of these methods are shown in Table 1.1. Corresponding error plots are demonstrated in Figure 1.2.

It can be seen that all six methods can approximate the solutions of the dual fuzzy equations. The neural network method is more suitable for solving these kinds of equations. The estimated errors of the neural network based algorithm are less than the other methods. The neural network method is faster and more robust when compared with the other methods. ∎

Example 1.4 The deformation of a solid cylindrical rod depends on the stiffness E, the forces on it F, the positions of the forces L, and the diameter of the rod d, see Figure 1.3. The positions are not exact, $L_1 = F(0.2, 0.3, 0.5, 0.6)$, $L_2 = F(0.4, 0.6, 0.7, 0.8)$, and $L_3 = F(0.4, 0.6, 0.7, 0.8)$. The area of the rod is $A = \frac{\pi}{4}d^2$. The external forces are a function of x, $F_1 = x^7$, $F_2 = x^6\sqrt{x}$,

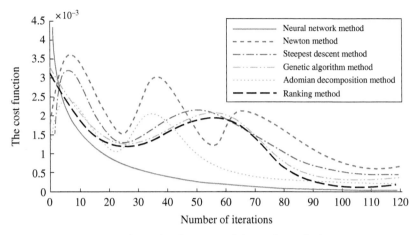

Approximation errors of six popular methods

Figure 1.2 Approximation errors of the six popular methods.

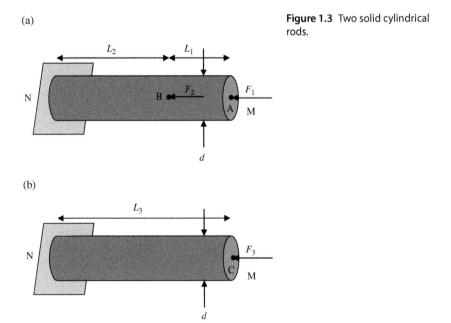

(a)

Figure 1.3 Two solid cylindrical rods.

(b)

and $F_3 = e^{2x}$. If the desired deformation at the point N is desired to be $N^* = F(0.000563, 0.000822, 0.001003, 0.001211)$, what is the quantity of the control force x?

Table 1.2 Approximation errors of the solid cylindrical rod.

k	Newton	Steepest descent	Genetic algorithm	Adomian decomposition	Ranking	Neural network
1	0.150	0.201	0.486	0.267	0.600	0.788
2	0.229	0.299	0.574	0.33	0.498	0.500
3	0.311	0.184	0.407	0.239	0.379	0.310
⋮	⋮	⋮	⋮	⋮	⋮	⋮
89	0.109	0.0801	0.0699	0.0594	0.0500	0.0098
90	0.0960	0.072	0.0600	0.0490	0.0411	0.0071

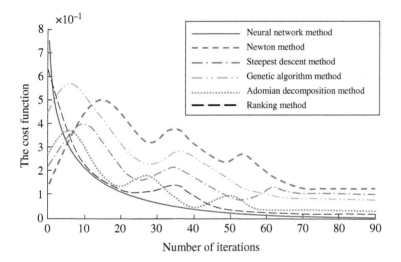

Approximation errors of six popular methods

Figure 1.4 Approximation error of the six popular methods.

The tension relation is [42],

$$\frac{L_1 F_1}{AE} \oplus \frac{L_2(F_1 + F_2)}{AE} = \frac{L_3 F_3}{AE} \oplus N^* \tag{1.35}$$

where $d = 0.02$ and $E = 70 \times 10^9$. The exact solution is $x = 4$. To approximate the solution, six popular methods are used: the Newton method, the steepest descent method, the genetic algorithm method, the Adomian decomposition method, the ranking method, and the neural network method. The errors of these methods are shown in Table 1.2. Neural network method is more robust than the other methods. Furthermore, the estimated error of the neural network is less when compared with other methods. Corresponding error plots are demonstrated in Figure 1.4.

1.5 Summary

In this chapter, some of the numerical methodologies are demonstrated as solutions of fuzzy equations and dual fuzzy equations. This review illustrates that the fuzzy roots of fuzzy equations can be obtained with different algorithms. However, in a few cases there exist no fuzzy roots in a particular fuzzy equation. The solution of the fuzzy polynomial by the ranking methodology is proposed for solving the fuzzy polynomial equation, which converts to a crisp system of polynomial equations, and therefore the system is easily solvable. For obtaining the real roots of the system in the case that there is no exact solution, iteration methodologies can be utilized for estimating the solution. Using a modified Adomian decomposition methodology, the fuzzy roots can be obtained by proposing fuzzy polynomials in parametric form and solving by the Adomian decomposition methodology. Obtaining a solution using the Newton methodology demonstrates that the fuzzy roots of fuzzy equations can be obtained at the initial step with high accuracy. By contrast, with the fuzzy neural network, the fuzzy root of the fuzzy equation can be obtained by proposing a learning algorithm. This review provides input for those showing interest in the field of fuzzy equations.

2

Fuzzy Differential Equations

2.1 Introduction

A review of the methodologies associated with the modeling and control of uncertain nonlinear systems is given due importance in this chapter. The basic criteria that highlight the work rely on the various patterns of techniques incorporated into the solutions of fuzzy differential equations (FDEs) that correspond to the controllability constraint related to fuzzy control. The solutions that are generated by these equations are considered to be the controllers. Currently, numerical techniques have emerged as superior techniques for solving these types of problems. Taking into consideration the modeling case as well as the control of uncertain nonlinear systems, the implementation of a neural networks technique has contributed to the complex method of dealing with the appropriate coefficients and solutions of fuzzy systems. In the current context, a few types of neural networks have been demonstrated, feed forward (static) and recurrent (dynamic), that are at par with least mean square and quasi-Newton learning techniques. In the context of this review, the applications and effectiveness of fuzzy control design techniques in the real world are reviewed.

2.2 Predictor–Corrector Method

The predictor–corrector methodology is broadly utilized in order to resolve initial value problems. In [27], three numerical methodologies for resolving fuzzy ordinary differential equations (ODEs) are proposed. These methodologies are Adams–Bashforth, Adams–Moulton and predictor–corrector. The predictor–corrector is extracted by blending the Adams–Bashforth and Adams–Moulton methodologies. The convergence and stability of the suggested methodologies are proved. Considering the convergence order of the Euler methodology which is $O(h)$ (as given in [121]), a higher order of convergence is achievable by utilizing the suggested methodologies in [27]. It should be mentioned that a predictor–corrector methodology of convergence

Modeling and Control of Uncertain Nonlinear Systems with Fuzzy Equations and Z-Number,
First Edition. Wen Yu and Raheleh Jafari.
© 2019 by The Institute of Electrical and Electronics Engineers, Inc. Published 2019 by John Wiley & Sons, Inc.

order $O(h^m)$ is utilized where the Adams–Bashforth m-step methodology and Adams–Moulton $(m - 1)$-step methodology are taken to be predictor and corrector, respectively. Following the ideas of [169], the suggested methodologies in [27] can resolve stiff problems.

In [37], it was shown that the exact solutions demonstrated in [27] are not solutions associated with FDEs and the correct exact solutions related to these problems are illustrated in [37]. Two characterization theorems are laid down for the solutions of FDEs that are employed for converting an FDE into a system of ODEs. The characterization theorems reveal the following investigation direction for FDEs. Convert the FDE into a system of ODEs, and then resolve the system of ODEs, as it is feasible to go back to the original FDE. This scheme can be employed with other differentiability concepts that include more natural attributes in comparison with the Hukuhara derivative.

In [114], a numerical solution concerned with hybrid FDEs is researched. An improved predictor–corrector methodology is selected and altered in order to resolve the hybrid FDEs on the basis of the Hukuhara derivative. The symbolic systems associated with the computer, Maple and Mathematica, are employed to carry out complex computations of algorithms. It is shown that the solutions extracted using the predictor–corrector methodology are more precise and well matched with the exact solutions.

In [34], an improved predictor–corrector method is presented in order to resolve FDEs under generalized differentiability. The generalized characterization theorem is used for converting an FDE into two ODE systems. The significance of transforming an FDE into a system of ODEs is that any numerical technique that is suitable for ODEs can be applied. The improved predictor–corrector three-step methodology can be generated to improved predictor–corrector m-step methodologies of convergence order $O(h^m)$.

In [25] an improved predictor–corrector method is suggested in order to resolve FDEs. The improved predictor–corrector method is generalized by combining an explicit three-step method and an implicit two-step method. The improved predictor–corrector three-step method is accurate and obtains a superior approximation. It can be generalized to improve predictor–corrector m-step methods.

The predictor–corrector technique is efficient since it utilizes information from previous steps. The drawback of the predictor–corrector technique is that the number of iterations is unknown. Furthermore, this technique is very difficult to program. As long as the solutions for sufficient points are defined, another technique such as the Adomian decomposition technique must be utilized.

2.3 Adomian Decomposition Method

In [156] the Adomian decomposition method is used for finding the fuzzy solution of homogeneous fuzzy partial differential equations (PDEs) with a specific fuzzy boundary and initial conditions. The Seikkala derivative is utilized for resolving fuzzy heat equations with a specific fuzzy boundary and initial conditions. The crisp form of the heat equation is resolved by utilizing the Adomian decomposition method. After that the solution is extended in fuzzy form as a Seikkala solution.

In [152] the Adomian decomposition method is implemented for finding the numerical solution of hybrid FDEs. This methodology considers the approximate solution of a nonlinear equation as an infinite series that generally converges to an accurate solution. A comparison between the approximation solutions and the exact solutions shows that the convergency is quite close.

In [11] the convergence of the Adomian decomposition technique is proved for an initial-value problem. The convergence rates of the Adomian decomposition technique are studied in the context of the nonlinear Schrödinger equation.

The great advantage of the Adomian decomposition technique is related to its application to all types of integral equations, linear or nonlinear, homogeneous or non-homogeneous, having constant coefficients or having variable coefficients. In addition, this technique is suitable for significantly reducing the size of the calculation process compared with traditional approaches. The drawback of this technique is that even though the series can be quickly convergent in a minute region, it has an extremely slow convergence rate in a broader region, and the truncated series solution is an imprecise solution in that region. There are other numerical techniques for solving FDEs, such as Euler technique, which is usually the next method investigated after the Adomian decomposition method. The Euler method is clear and simple to understand.

2.4 Euler Method

In [121], the FDE is substituted by its parametric form. The classical Euler technique is implemented for resolving a novel system that contains two classical ODEs with initial conditions. The capability of the technique is demonstrated by resolving several linear as well as nonlinear first order FDEs.

In [142], a linear first order FDE is analyzed by employing a strongly generalized differentiability approach. This work is based on the generalized characterization theorem in which the FDE is substituted with its equivalent systems, and so for estimating the two fuzzy solutions two ODE systems are resolved by using the generalized Euler approximation technique that includes four classic ODEs having initial conditions. Furthermore, an error analysis of the generalized Euler technique, which assures pointwise convergency, is laid down. The significance of transforming a FDE into a system of ODEs is based on the fact that any numerical technique that is appropriate for ODEs can be employed.

In [18], a novel fuzzy version of Euler's methodology is suggested in order to solve differential equations with fuzzy initial values. The suggested methodology relies on Zadeh's extension principle in order to reformulate the classical Euler methodology, which considers the dependency problem that is generated in a fuzzy setting. This problem is regularly avoided in numerical methodologies mentioned in the literature in order to resolve differential equations having fuzzy initial values. This paper has positive attributes when compared with the traditional fuzzy version of Euler's methodology. At par with [45], taking into consideration the non-dependency problem associated with fuzzy calculation will result in the repetition of some numerical calculations. Therefore, there prevails expected errors, and in the last phase the errors may generate estimations that are broader in comparison with the correction. This is authentic as the initial results carried out in [19] have demonstrated that the solution of FDEs extracted by utilizing the suggested methodology in [121] has an overestimation in calculation.

In [143], convergence as well as the stability of Euler the technique in order to solve fuzzy stochastic differential equations under the generalized differentiability concept are investigated.

In [186], two improvised Euler type methodologies, named as the max-min improved Euler methodology and the average improved Euler methodology, are suggested for extracting numerical solution of linear as well as nonlinear ODEs at par with the fuzzy initial condition. In this paper all the possible blends of lower as well as upper bounds concerned with the variable are considered and then resolved by the suggested methodologies. Also, an exact method is laid down.

In [185], the numerical solution associated with linear, nonlinear, and a system of ODEs with a fuzzy initial condition is researched. Two Euler type methodologies, namely the max-min Euler methodology and the average Euler methodology, are laid down for extracting a numerical solution related to the FDEs. Several investigators have considered the left and right bounds of the variables in the differential equations. In this paper, the investigators constructed the methodologies by taking into account all possible combinations of lower as well as upper bounds of the variable. The solution extracted by the max-min Euler methodology very closely matches with the outcomes extracted by [121] and an exact solution.

For many higher order systems, it is very difficult to make the Euler approximation effective. The Euler methodology is not very accurate and stable. High order Taylor series methods are more accurate than the Euler method.

2.5   Taylor Method

In [4], an approach on the basis of the second Taylor technique is illustrated in order to resolve linear as well as nonlinear FDEs. The convergence order of the Euler technique in [121] is $O(h)$, whereas the convergence order in [4] is $O(h^2)$. The better solutions are extracted by [4].

In [5] the Taylor method of order p is utilized for solving FDEs. The algorithm is explained by resolving some linear and nonlinear fuzzy Cauchy problems. The convergence order of the Taylor method is $O(h^p)$.

The drawback of the Taylor series technique is the computation of higher derivatives, in which increasing the order the calculation process becomes increasingly complicated. However, the Runge–Kutta method is generally considered to be the most effective one-step technique.

2.6   Runge–Kutta Method

In [151] an effective s stage Runge–Kutta technique is employed for extracting the numerical solution of the FDE. In this paper, the Runge–Kutta method is applied for a more generalized category of problems and a convergence definition as well as error definitions are given, at par with FDE theory. Furthermore, convergence related to s stage Runge–Kutta methods is analyzed. This technique, when compared with the developed Euler technique, performs in a superior manner. Although the Euler technique is suitable, it is embedded with the disadvantage that, when analyzing the convergence of the Euler technique [121], the authors generally investigate the convergence of the ODE system, which occurs while resolving numerically.

In [6], a numerical algorithm in order to resolve linear as well as nonlinear fuzzy ODEs on the basis of Seikkala's derivative of fuzzy process is investigated. A numerical methodology on the basis of a four-stage order Runge–Kutta technique is stated and is carried out using a complete error analysis. However, their work generalizes the same problems as those mentioned in [121], and also relies totally on four-stage methodologies.

In [105], a numerical algorithm is suggested in order to solve linear as well as nonlinear fuzzy ODEs on the basis of the Seikkala derivative of a fuzzy process. A numerical technique on the basis of the Runge–Kutta methodology of order five is elaborately investigated and this is carried out by going through an analysis of complete error. This technique with $O(h^5)$ outperforms the improved Euler technique with $O(h^2)$.

In [105], a numerical solution for nth order FDEs on the basis of the Seikkala derivative having an initial value problem is investigated. The Runge–Kutta–Nystrom technique is employed for extracting the numerical solution of this problem, and the convergency and stability of the method is validated. In this methodology the nth order FDE is transformed to a fuzzy system that can be resolved by utilizing the Runge–Kutta–Nystrom methodology.

A family of extended Runge–Kutta-like formulae are implemented in [73]. The formula reveals the utilization of first derivatives f'. This approach uses an extended Runge–Kutta methodology involving a local truncation error of order five with only four evaluations of both f as well as f' to approximate the local error in an extended Runge–Kutta methodology having order four per step. The positive attribute of this method is that just four evaluations of both f as well as f' are needed per step, whilst arbitrary classical Runge–Kutta methodologies of orders three and four employed in combination require six evaluations of f per step. In [73], a numerical algorithm in order to resolve the fuzzy first order initial value problem on the basis of extended Runge–Kutta-like formulae of order four is implemented. In this paper the extended Runge–Kutta-like formula is employed for enhancing the order of preciseness related to the solutions by evaluating both f and f' instead of only evaluating f.

In [109], a numerical algorithm in order to solve FDEs on the basis of Seikkala's derivative of a fuzzy process is suggested. A numerical technique based on a Runge–Kutta–Nystrom technique of order three is employed for solving the initial value problem, and it is also illustrated that this methodology is superior in comparison with the Euler method by considering the convergence order of the Euler methodology ($O(h)$) as well as the Runge–Kutta–Nystrom methodology ($O(h^3)$).

The major advantages of the Runge–Kutta technique is that it is easy to apply. The main drawbacks of the Runge–Kutta technique is that it needs more computer time when compared with multi-step techniques, and it does not easily yield desirable global approximations of the truncation error. The finite difference method is comparatively simple as well as computationally rapid.

2.7   Finite Difference Method

Finite difference methodologies illustrate functions as discrete values across a grid and estimate their derivatives as differences between points on the grid. In [68] a numerical methodology for solving the fuzzy heat equation is laid down. A difference approach is taken into account for the one dimensional heat equation. Additionally, the necessary conditions for stability of the suggested approach is illustrated. The suggested difference methodology associated with the example mentioned in [68] is tested when the exact solution is known. In this example the Hausdorff distance between the exact solution and the estimated solution is extracted.

In [119], the finite difference methodology for resolving differential equations is proposed. Furthermore, finite difference estimations to higher order derivatives are suggested.

In [139], an implicit finite difference methodology for resolving fuzzy PDEs is discussed. The stability related to this methodology is demonstrated and the parabolic equation is resolved with this approach.

In [17], a numerical methodology on the basis of the Seikkala derivative for resolving fuzzy PDEs is taken into account. A difference methodology for solving fuzzy PDEs, called a fuzzy hyperbolic equation as well as a fuzzy parabolic equation, is illustrated, and the stability associated with this methodology is analyzed and conditions for stability are supplied. If all terms of fuzzy PDEs belong to a fuzzy set, then the solutions of fuzzy PDEs prevail, which are concluded from the numerical values. Examples demonstrate that the Hausdorff distance between the exact solution and the approximate solution is minute.

In [87], it is shown how to induce boundary conditions into finite difference methodologies so the resulting estimations copy the identities between the differential operators of vector as well as tensor calculus. The scheme is valid for a wide class of PDEs, and it is stated for Poisson's equation with Dirichlet, Neumann as well as Robin boundary conditions. These estimations retain the most important properties of original differential problems. In particular, the discrete estimation is symmetric and positive definite. The properties associated with the discrete operators make it feasible to utilize efficient iteration methodologies in order to resolve a system of linear equations.

In [24], difference methodologies based on the Seikkala derivative for resolving linear and nonlinear fuzzy PDEs are considered. A fuzzy reachable set can be estimated using the suggested methodologies including a complete error analysis. The methodologies are demonstrated by resolving three types of fuzzy PDEs, namely the finite difference Poisson equation (the difference methodology utilizing the Taylor series), the backward difference heat equation (the Crank–Nicolson methodology), and the finite difference wave equation.

In [112], an implicit finite difference methodology is discussed in order to solve fuzzy PDEs on the basis of the Siekkala derivative. Furthermore, the stability of the methodology is laid down.

The major disadvantage of the finite difference technique is its flexibility. The finite difference technique may generate spurious oscillations as well as negative solutions due to truncation errors, and also may become unstable. The differential transform method can be utilized directly on nonlinear initial value problems without the need for linearization and discretization, therefore it is not influenced by the errors of discretization.

2.8 Differential Transform Method

In [128], a two dimensional differential transform methodology of fixed grid size is employed to obtain approximate solutions related to fuzzy PDEs. In addition, an adaptive grid size mechanism based on the fixed grid size methodology is laid down. The suggested methodologies generate a Taylor series expansion solution for the domain between any adjacent grid points. This methodology is effective for extracting exact and approximate solutions of linear ordinary, nonlinear ordinary, and fuzzy PDEs. The differential transform methodology provides an analytical solution in the polynomial form. The differential transform methodology converts the PDEs as well as related initial conditions into a recurrence equation, which ultimately results in the solution of a system related to algebraic equations as coefficients of a power series solution. It is different from the conventional high order Taylor series methodology that needs symbolic calculation of the essential derivatives of the data functions, and is also computationally costly for high order. The fuzzy differential transform methodology evaluates the estimating solution by utilizing a finite Taylor series. The fuzzy differential transform methodology is barred from evaluating the derivatives symbolically. It computes the relative derivatives by using an iteration technique stated by the fuzzy transformed equations extracted from the original equations utilizing fuzzy differential transformation.

In [29], an extension of the differential transformation methodology in order to resolve the FDE is proposed. Different types of exact, approximate, and purely numerical methodologies are given to extract the solution related to a fuzzy initial value problem. The investigators have generally gone through the methodologies on the basis of the Hukuhara derivative. However, in some situations this concept undergoes a severe drawback. The solution includes a property that $\dim(x(t))$ is non-decreasing in t, i.e. the solution is irreversible in possibilistic terms. Hence, this explanation is not an appropriate generalization of the associated crisp case. It is considered that this problem is the outcome of fuzzification related to the the derivative employed in the development of the FDE. Additionally, most known methodologies of resolving FDEs are computationally intense, since they are trial-and-error in nature, or require complex symbolic calculations. In [29], the authors have solved these difficulties by utilizing a more simple definition of the derivative, at par with the fuzzy mappings, extending the class of differentiable fuzzy mappings that are laid down in [38, 39], and then utilized differential transformation methodology for solving FDEs.

In [36], a generalization of the differential transformation methodology in order to solve the fuzzy PDE by utilizing the strongly generalized differentiability approach is researched.

In [128], a differential transform technique is different from the conventional high order Taylor series technique, which needs symbolic calculation of essential derivatives of the data function, and is also computationally costly for higher orders. In this paper, the fuzzy differential transform method is suggested for resolving fuzzy PDEs. A few examples are analyzed by utilizing the fuzzy differential transform method and the outcomes exhibit significant performance.

In [76], the differential transformation method is applied for finding the numerical solution for hybrid FDEs. This technique considers the approximate solution of a nonlinear equation as an infinite series that is generally convergent to the exact solution. It is based on the Taylor series expansion that generates the solution in the form of a polynomial series solution by utilizing an iterative procedure. In this technique, the preciseness of the extracted solution can be improved by taking more terms in the solution.

Whereas the differential transform method generates a truncated Taylor series solution, so it does not represent a desirable estimation in an extensive domain. Due to the superior estimation abilities of neural networks, the estimated solution for the FDE is extremely near to the exact solution.

2.9 Neural Network Method

In [116], a technique in order to resolve both ODEs and PDEs is presented, and is dependent on the function approximation abilities of feed forward neural networks. This technique results in the development of a solution presented in a differentiable and closed analytic form. This form applies a feed forward neural network as the basic estimation element and its parameters (weights and biases) are adjusted to diminish a suitable error function. In order to train the network, optimization methodologies are implemented that need the calculation of the gradient error considering the network parameters. In the suggested methodology the model function is presented as the sum of two terms. The first term meets the initial/boundary conditions, and also does not include adjustable parameters. The second term includes a feed forward neural network to be trained in order to meet the needs of the differential equation. The implementation of a neural architecture sums up several attractive features of the technique:

1) The solution through an artificial neural network is differentiable, having closed analytic form, which is easily utilized in any subsequent computation. Most other methodologies suggest a discrete solution (viz the predictor–corrector or Runge–Kutta methodologies) or a solution of limited differentiability (viz finite elements).

2) The implementation of neural networks supplies a solution with highly superior generalized attributes. Results compared with the finite element methodology that is depicted in this work describe this point vividly.

3) The required number of model parameters is very much less than any other method and hence compact solution models are extracted with low demand criteria on memory space.

4) The technique is simple and can be implemented on ODEs, systems of ODEs, and PDEs stated on orthogonal box boundaries. Furthermore, the process is in an advancement in dealing with the case of irregular (arbitrarily shaped) boundaries.

5) The technique can be tested in hardware, utilizing neuro processors, and proposes an opportunity to handle real time complex differential equation problems that occur in several engineering applications.

6) The technique can be effectively imposed on parallel architectures.

This technique is simple and can be employed on both ODEs and PDEs by developing a suitable form of the trial solution. The technique displays superior generalization performance as the deviation at the test points is in no case more than the maximum deviation at the training points. This is in contrast with the finite element technique in the case that the deviation at the testing points is extremely high in comparison with the deviation at the training points.

In [1], a technique is introduced in order to resolve PDEs with boundary and initial conditions by utilizing neural networks. An evolutionary algorithm is employed for training the networks. The outcomes of implementing the methodology on one dimensional and two dimensional problems are highly superior and convincing.

However, the principle concept is that evolutionary algorithms uncover all regions of the solution space as well as exploit favorable areas via implementing recombination, mutation, selection, and reinsertion operations on the individuals of a population. Usually all approaches that are employed on a population cause the development of a new generation. In [1] the researchers worked considering a single population. In [158], it is illustrated that single population evolutionary algorithms are strong and perform well on a broad variety of problems. Nevertheless outcomes are highly effective when working with multiple sub-populations in lieu of just a single population.

In [86] a modified technique is proposed in order to obtain the numerical solutions of fuzzy PDEs by utilizing fuzzy artificial neural networks. Utilizing a modified fuzzy neural network ensures that the training points are selected over an open interval without training the network in the range of the first and end points. This novel technique is based on substituting each x in the training set (where $x \in [a, b]$) by the polynomial $Q(x) = \epsilon(x + 1)$ in such a manner that $Q(x) \in (a, b)$, by selecting an appropriate $\epsilon \in (0, 1)$. Also, it can be suggested that the proposed methodology can deal efficiently with all types of fuzzy PDEs as well as generating a precise estimated solution for all the domain and not only for the training set. Hence, one can utilize the interpolation methodologies (called the curve fitting methodology) in order to obtain the estimated solution at points in the midst of the training points or at points outside the training set.

In [135], a new hybrid technique based on a learning algorithm of fuzzy neural networks for extracting the solution to a differential equation with a fuzzy initial value is demonstrated. The model obtains the estimated solution to the FDE inside of its domain in the neighborhood of the fuzzy initial point. One deficiency of fully fuzzy neural networks along with fuzzy connection weights is the long calculation time. Another deficiency is that the learning algorithm is complex. In order to minimize the complexity of the learning algorithm, in [135] a partially fuzzy neural network architecture is demonstrated by taking into account that the connection weights to the output unit are fuzzy numbers, whereas the connection weights and biases to hidden units are real numbers. By using different learning algorithms, different simulation outcomes can be achieved. For example, some global learning algorithms called genetic algorithms can train non-fuzzy connection weights better than the back propagation type learning algorithm for fuzzy mappings of triangular shape fuzzy numbers having incremental fuzziness. Also the utilization of more general network architectures causes the back propagation type learning algorithm to be more complex.

In [136], a technique for estimating the solution to a second order FDE is suggested by utilizing a fuzzy neural network on the basis of back propagation type learning algorithms. The employment of more simplified network architectures causes the back propagation type learning algorithm to be highly complex. In [136], notable simulation results from a partially fuzzy neural network are demonstrated but the researchers did not elaborate their learning algorithm for neural networks with more than three layers.

In [84], a novel technique on the basis of a learning algorithm associated with fuzzy neural networks as well as a Taylor series is laid down for extracting the numerical solution of FDEs. A fuzzy neural network on the basis of the semi-Taylor series (in relation to the function e^x) for the first (and second) order FDE is utilized. It is possible to use the same approach for solving high order FDEs as well as fuzzy PDEs. A fuzzy trial solution related to the fuzzy initial value problem is presented as an addition of two parts. The primary part meets the fuzzy initial condition, includes a Taylor series, and contains no fuzzy adjustable parameters. The secondary part includes a feed forward fuzzy neural network having fuzzy adjustable parameters (the fuzzy weights). Therefore, by development, the fuzzy primary condition is met and the training of the fuzzy network is carried out in order to meet the needs of the FDE. The preciseness of this technique relies on the Taylor series that is selected for the trial solution. This selection is not distinct, hence the preciseness is different from one problem to another problem. The suggested technique gives more precise estimations. Superior outcomes will be possible if more neurons or more training

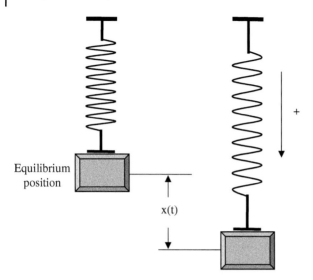

Figure 2.1 Vibration mass.

points are used. In addition, after resolving a FDE the solution is achievable at any arbitrary point in the training interval (even in the midst of training points).

In order to compare the efficiency of the numerical methods to approximate the solution to FDEs the below examples are demonstrated.

Example 2.1 The vibration mass system displayed in Figure 2.1 is modeled by

$$\frac{\mathrm{d}}{\mathrm{d}t}u(t) = \frac{k}{m}x(t), \qquad u(t) = \frac{\mathrm{d}}{\mathrm{d}t}x(t) \tag{2.1}$$

where the spring constant is considered to be $k = 1$, and the mass is $m = (0.73, 1.123)$. If the initial position is taken to be $x(0) = (0.73 + 0.23\alpha, 1.123 - 0.123\alpha)$, $\alpha \in [0, 1]$, so the exact solutions of Equation (2.1) are [78]

$$x(t, \alpha) = [(0.73 + 0.23\alpha)e^t, (1.123 - 0.123\alpha)e^t] \tag{2.2}$$

where $t \in [0, 1]$.

To approximate the solution (2.2), six popular methods are used: the predictor–corrector method, the Adomian decomposition method, the Euler method, the Taylor method, the Runge–Kutta method, and the neural network method. The errors of these methods are shown in Table 2.1. Corresponding solution plots are displayed in Figure 2.2.

All six methods are suitable for resolving the FDEs. The learning procedure of the neural network method is much quicker than the other methods. Also, the robustness of neural network method is better when compared with the other methods. ∎

Table 2.1 Approximation errors of a vibration mass system.

α	Predictor–corrector	Adomian decomposition	Euler	Taylor	Runge–Kutta	Neural network
0	[0.2061,0.4385]	[0.0927,0.1407]	[0.0751,0.1209]	[0.0604,0.1088]	[0.0407,0.0889]	[0.0209,0.0606]
0.2	[0.2232,0.4572]	[0.1032,0.1509]	[0.0862,0.1308]	[0.0701,0.1191]	[0.0609,0.1092]	[0.0308,0.0704]
0.4	[0.1959,0.4278]	[0.0831,0.1309]	[0.0649,0.1102]	[0.0511,0.0993]	[0.0211,0.0692]	[0.0102,0.0501]
0.6	[0.1859,0.4178]	[0.0724,0.1201]	[0.0548,0.1002]	[0.0403,0.0880]	[0.0211,0.0691]	[0.0011,0.0409]
0.8	[0.2472,0.4791]	[0.1231,0.1711]	[0.1162,0.1512]	[0.1009,0.1493]	[0.0712,0.1192]	[0.0512,0.0913]
1	[0.2573,0.2573]	[0.1431,0.1431]	[0.1254,0.1254]	[0.1104,0.1104]	[0.0806,0.0806]	[0.0602,0.0602]

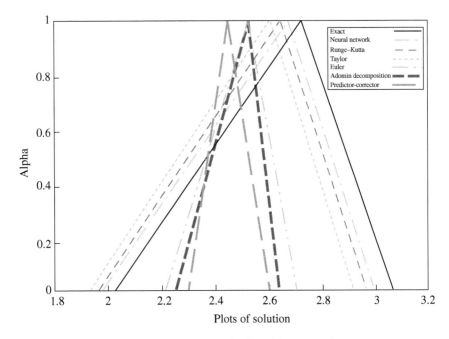

Figure 2.2 Comparison plot of six popular methods and the exact solution.

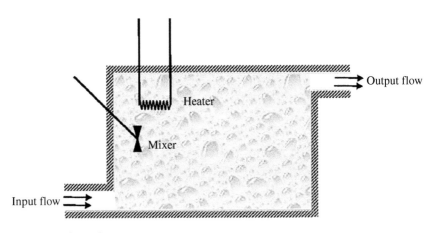

Figure 2.3 Thermal system.

Table 2.2 Approximation errors of the heating system.

α	Predictor–corrector	Adomian decomposition	Euler	Taylor	Runge–Kutta	Neural network
0	[0.2275,0.4507]	[0.1196,0.1623]	[0.0902,0.1423]	[0.0812,0.1295]	[0.0695,0.1507]	[0.0423,0.0812]
0.2	[0.2159,0.4407]	[0.1006,0.1553]	[0.0839,0.1367]	[0.0742,0.1883]	[0.0473,0.0863]	[0.0381,0.0702]
0.4	[0.2419,0.4718]	[0.1228,0.1713]	[0.0112,0.1545]	[0.0954,0.1303]	[0.0864,0.1092385]	[0.0519,0.0901]
0.6	[0.2613,0.4962]	[0.1486,0.1923]	[0.1385,0.1773]	[0.1231,0.1684]	[0.0933,0.1401]	[0.0734,0.1578]
0.8	[0.2009,0.4319]	[0.0933,0.1441]	[0.0774,0.1295]	[0.0692,0.1063]	[0.0427,0.0879]	[0.0201,0.0635]
1	[0.2785,0.2785]	[0.1633,0.1633]	[0.1448,0.1448]	[0.1327,0.1327]	[0.1003,0.1003]	[0.0801,0.0801]

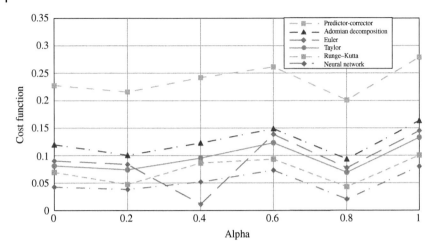

Figure 2.4 The lower bounds of absolute errors.

Example 2.2   A tank with a heating system is shown in Figure 2.3, where $R = 0.5$ and the thermal capacitance is $C = 2$. The temperature is x. The model is [157],

$$\frac{\mathrm{d}}{\mathrm{d}t}x(t) = -\frac{1}{RC}x(t) \qquad (2.3)$$

where $t \in [0, 1]$ and x is the amount of sinking in each moment. If the initial position is $x(0) = (\alpha - 1, 1 - \alpha)$ and $\alpha \in [0, 1]$, then the exact solutions of (2.3) are

$$x(t, \alpha) = [(\alpha - 1)e^t, (1 - \alpha)e^t]. \qquad (2.4)$$

To approximate the solution (2.4), six popular methods are used: the predictor–corrector method, the Adomian decomposition method, the Euler method, the Taylor method, the Runge–Kutta method, and the neural network method. The errors of these methods are shown in Table 2.2. The lower and upper bounds of absolute errors are shown in Figure 2.4 and Figure 2.5 respectively. The approximation errors of the neural network method are smaller than the other methods. ∎

2.10   Summary

A review of the methodologies for modeling and control of uncertain nonlinear systems has been given in this chapter. The application of the FDEs is in direct connection with the nonlinear modeling and control. It is feasible for these equations to apply directly for nonlinear control. Fuzzy control

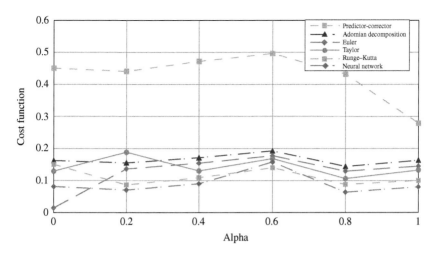

Figure 2.5 The upper bounds of absolute errors.

with FDEs requires the solution of the FDEs. Several approaches such as the predictor–corrector method, the Adomian decomposition method, the Euler method, the Taylor method, the Runge–Kutta method, The finite difference method, the differential transform method, and the neural network method are given.

3

Modeling and Control Using Fuzzy Equations

In this chapter, more general dual fuzzy equations [192] are discussed. For modeling, the fuzzy equation is initially transformed into a neural network. Then the normal gradient descent method is modified to train the fuzzy coefficients. With this modification, normal neural modeling methods are applied to uncertainty nonlinear system modeling with fuzzy equations. The approximation theories for crisp models are extended into fuzzy cases. The upper bounds of the modeling errors with fuzzy equations are estimated. Some simulation results are provided to show the performance and effectiveness of the fuzzy control and modeling design methods with neural networks.

3.1   Fuzzy Modeling with Fuzzy Equations

Normal fuzzy equations have fuzzy numbers only on one side of the equation. However, dual fuzzy equations have fuzzy numbers on both sides of the equation. Since the fuzzy numbers cannot be moved between the sides of the equation [108], dual fuzzy equations are more general and difficult. Initially, the existence of solutions to dual fuzzy equations is discussed. This corresponds to the controllability problem of fuzzy control [58]. Then two methods to approximate the solutions of the dual fuzzy equations are provided. They are a controller design process.

Consider the following unknown discrete time nonlinear system

$$\overline{x}_{k+1} = \overline{f}\left[\overline{x}_k, u_k\right], \quad y_k = \overline{g}\left[\overline{x}_k\right] \tag{3.1}$$

where $u_k \in R^u$ is the input vector, $\overline{x}_k \in R^l$ is an internal state vector, and $y_k \in R^m$ is the output vector. $\overline{f}$ and $\overline{g}$ are general nonlinear smooth functions $\overline{f}, \overline{g} \in C^\infty$. Denote $Y_k = \left[y_{k+1}^T, y_k^T, \ldots\right]^T$ and $U_k = \left[u_{k+1}^T, u_k^T, \ldots\right]^T$. If $\frac{\partial Y}{\partial x}$ is non-singular at $\overline{x} = 0$ and $U = 0$, then this leads to the following model

$$y_k = \Phi\left[y_{k-1}^T, y_{k-2}^T, \ldots u_k^T, u_{k-1}^T, \ldots\right] \tag{3.2}$$

Modeling and Control of Uncertain Nonlinear Systems with Fuzzy Equations and Z-Number, First Edition. Wen Yu and Raheleh Jafari.
© 2019 by The Institute of Electrical and Electronics Engineers, Inc. Published 2019 by John Wiley & Sons, Inc.

where $\Phi(\cdot)$ is an unknown nonlinear difference equation representing the plant dynamics, and u_k and y_k are measurable scalar input and output, respectively. The nonlinear system (3.2) is a NARMA model. The input of the nonlinear system can be demonstrated as

$$x_k = \left[y_{k-1}^T, y_{k-2}^T, \cdots u_k^T, u_{k-1}^T, \cdots \right]^T \tag{3.3}$$

and the output as y_k.

Many nonlinear systems as in (3.2) can be rewritten as the following linear-in-parameter model,

$$y_k = \sum_{j=0}^{n} a_j x_k^j \tag{3.4}$$

or

$$y_k = \sum_{i=1}^{n} a_i f_i(x_k) \tag{3.5}$$

or

$$y_k + \sum_{i=1}^{m} b_i g_i(x_k) = \sum_{i=1}^{n} a_i f_i(x_k) \tag{3.6}$$

where a_j, a_i, and b_i are linear parameters, and $x_k^j, f_i(x_k)$, and $g_i(x_k)$ are nonlinear functions. The variables of these functions are measurable input and output.

A popular example of this kind of model is the robot manipulator [176]

$$M(q)\ddot{q} + C(q,\dot{q})\dot{q} + B\dot{q} + g(q) = \tau. \tag{3.7}$$

Equation (3.7) can be rewritten as

$$\sum_{i=1}^{n} Y_i(q,\dot{q},\ddot{q})\theta_i = \tau. \tag{3.8}$$

To identify or control the linear-in-parameter system (3.4)–(3.6), or (3.8) the normal least square or adaptive methods can be applied directly.

Here the uncertain nonlinear systems can be considered, i.e. the parameters a_j, a_i, b_i, and θ_i are not fixed (not crisp). They are uncertain in the sense of fuzzy logic. The uncertain nonlinear systems are modeled by linear-in-parameter models with fuzzy parameters. These models are called fuzzy equations. The special case of fuzzy equations is the fuzzy polynomial interpolation. Before introducing fuzzy equations and fuzzy polynomial interpolation, the following definitions are required [197].

Definition 3.1 (Fuzzy number) *A fuzzy number u is a function $u \in E : R \rightarrow [0,1]$, such that, (1) u is normal, (there exists $x_0 \in R$ such that $u(x_0) = 1$; (2) u is convex, $u(\lambda x + (1-\lambda)y) \geq \min\{u(x), u(y)\}, \forall x, y \in R, \forall \lambda \in [0,1]$; (3) u is upper*

semi-continuous on R, i.e. $u(x) \le u(x_0) + \varepsilon, \forall x \in N(x_0), \forall x_0 \in R, \forall \varepsilon > 0, N(x_0)$
is a neighborhood; (4) the set $u^+ = \{x \in R, u(x) > 0\}$ is compact.

Membership functions are used to express the fuzzy number. The most popular membership functions are the triangular function

$$u(x) = F(a, b, c) = \begin{cases} \frac{x-a}{b-a} & a \le x \le b \\ \frac{c-x}{c-b} & b \le x \le c \end{cases} \qquad (3.9)$$

otherwise $u(x) = 0$ and the trapezoidal function

$$u(x) = F(a, b, c, d) = \begin{cases} \frac{x-a}{b-a} & a \le x \le b \\ \frac{d-x}{d-c} & c \le x \le d \\ 1 & b \le x \le c \end{cases} \qquad (3.10)$$

otherwise $u(x) = 0$.

Similarly, with crisp numbers, the fuzzy number u also has four basic operations: $\oplus, \ominus, \odot$, and $\oslash$. They represent the operations: sum, subtract, multiply, and multiply by a crisp number.

The dimension of x in the fuzzy number u depends on the membership function, for example (3.9) has three variables and (3.10) has four variables. In order to define consistency operations, at first an α-level operation is applied to the fuzzy number.

Definition 3.2 (α-level)   *The α-level of a fuzzy number u is defined as*

$$[u]^\alpha = \{x \in R : u(x) \ge \alpha\} \qquad (3.11)$$

where $0 < \alpha \le 1, u \in E$.

So $[u]^0 = u^+ = \{x \in R, u(x) > 0\}$. Because $\alpha \in [0, 1]$, $[u]^\alpha$ is bounded as $\underline{u}^\alpha \le [u]^\alpha \le \overline{u}^\alpha$. The α-level of u between $\underline{u}^\alpha$ and $\overline{u}^\alpha$ is defined as

$$[u]^\alpha = A\left(\underline{u}^\alpha, \overline{u}^\alpha\right). \qquad (3.12)$$

Let $u, v \in E$, $\lambda \in R$, the following fuzzy operations are defined. $\underline{u}^\alpha$ and $\overline{u}^\alpha$ are the function of $\alpha.\underline{u}^\alpha = d_M(\alpha), \overline{u}^\alpha = d_U(\alpha), \alpha \in [0, 1]$ are defined.

Definition 3.3 (Lipschitz constant)   [141] *The Lipschitz constant H of a fuzzy number $u \in E$ is*

$$\left|d_M(\alpha_1) - d_M(\alpha_2)\right| \le H\left|\alpha_1 - \alpha_2\right|$$
$$rmor \quad \left|d_U(\alpha_1) - d_U(\alpha_2)\right| \le H\left|\alpha_1 - \alpha_2\right| \qquad (3.13)$$

Definition 3.4 (Fuzzy operations)   [190] *Sum,*

$$[u \oplus v]^\alpha = [u]^\alpha + [v]^\alpha = [\underline{u}^\alpha + \underline{v}^\alpha, \overline{u}^\alpha + \overline{v}^\alpha]. \qquad (3.14)$$

Subtract,

$$[u \ominus v]^\alpha = [u]^\alpha - [v]^\alpha = [\underline{u}^\alpha - \underline{v}^\alpha, \overline{u}^\alpha - \overline{v}^\alpha]. \tag{3.15}$$

Multiply,

$$\underline{w}^\alpha \le [u \odot v]^\alpha \le \overline{w}^\alpha \quad \text{or} \quad [u \odot v]^\alpha = A\left(\underline{w}^\alpha, \overline{w}^\alpha\right) \tag{3.16}$$

where $\underline{w}^\alpha = \underline{u}^\alpha \underline{v}^1 + \underline{u}^1 \underline{v}^\alpha - \underline{u}^1 \underline{v}^1$, $\overline{w}^\alpha = \overline{u}^{\alpha} \overline{v}^{1} + \overline{u}^{1} \overline{v}^{\alpha} - \overline{u}^{1} \overline{v}^{1}$, $\alpha \in [0, 1]$. *It is a cross product of two fuzzy numbers.*

Multiply by a crisp number: for arbitrary crisp real positive number τ,

$$- \tau \underline{u}^\alpha \le [u]^\alpha \mathcal{O} \tau \le -\tau \overline{u}^\alpha$$

$$[u]^\alpha \mathcal{O} \tau = A\left(-\tau \underline{u}^\alpha, -\tau \overline{u}^\alpha\right). \tag{3.17}$$

Obviously, the following properties are held: the scalar multiplication: $\alpha \in [0, 1]$

$$[\lambda u]^\alpha = \lambda [u]^\alpha = \begin{cases} A\left(\lambda \underline{u}^\alpha, \lambda \overline{u}^\alpha\right) & \lambda \ge 0 \\ A\left(\lambda \overline{u}^\alpha, \lambda \underline{u}^\alpha\right) & \lambda < 0 \end{cases}. \tag{3.18}$$

$$\ominus u = (-1)u, \quad u \in E \tag{3.19}$$

Definition 3.5 (Dot product) [35] *The dot product of two fuzzy variables u and v is*

$$(u.v)^\alpha = A\left(\begin{array}{c} \min\{\underline{u}^\alpha \underline{v}^\alpha, \underline{u}^\alpha \overline{v}^\alpha, \overline{u}^\alpha \underline{v}^\alpha, \overline{u}^\alpha \overline{v}^\alpha\} \\ \max\{\underline{u}^\alpha \underline{v}^\alpha, \underline{u}^\alpha \overline{v}^\alpha, \overline{u}^\alpha \underline{v}^\alpha, \overline{u}^\alpha \overline{v}^\alpha\} \end{array}\right). \tag{3.20}$$

Definition 3.6 (Distance) *The distance between the fuzzy numbers u and v is*

$$d(u, v) = \sup_{0 \le \alpha \le 1} \{\max\left(|\underline{u}^\alpha - \underline{v}^\alpha|, |\overline{u}^\alpha - \overline{v}^\alpha|\right)\}. \tag{3.21}$$

Definition 3.7 (Absolute value) [8] *Absolute value of a triangular fuzzy number* $u(x) = F(a, b, c)$ *is*

$$|u(x)| = |a| + |b| + |c|. \tag{3.22}$$

Definition 3.8 (Positive) *A fuzzy number* $u \in E$ *is said to be positive if* $\underline{u}^1 \ge 0$ *and negative if* $\overline{u}^1 \le 0$.

Clearly, if u is positive and v is negative then $u \odot v = \ominus(u \odot (\ominus v))$ is a negative fuzzy number. If u is negative and v is positive then $u \odot v = \ominus((\ominus u) \odot v)$ is a negative fuzzy number. If u and v are negative then $u \odot v = (\ominus u) \odot (\ominus v)$ is a positive fuzzy number.

If u is positive and v is negative:

$$(u \odot v)^\alpha = A \left(\frac{\overline{u}^\alpha v^1 + \overline{u}^1 v^\alpha}{-\underline{u}^1 \underline{v}^1, \underline{u}^\alpha \overline{v}^1 + \underline{u}^1 \overline{v}^\alpha - \underline{u}^1 \overline{v}^1} \right). \tag{3.23}$$

If u is negative and v is positive:

$$(u \odot v)^\alpha = A \left(\frac{u^\alpha \overline{v}^1 + u^1 \overline{v}^\alpha}{-\underline{u}^1 \underline{v}^1, \overline{u}^\alpha \underline{v}^1 + \overline{u}^1 \underline{v}^\alpha - \overline{u}^1 \underline{v}^1} \right). \tag{3.24}$$

If u and v are negative:

$$(u \odot v)^\alpha = A \left(\frac{\overline{u}^\alpha \overline{v}^1 + \overline{u}^1 \overline{v}^\alpha}{-\overline{u}^1 \overline{v}^1, \underline{u}^\alpha \underline{v}^1 + \underline{u}^1 \underline{v}^\alpha - \underline{u}^1 \underline{v}^1} \right). \tag{3.25}$$

When the parameters in the linear-in-parameter model (3.4), (3.5), or (3.6) are fuzzy numbers, (3.4), (3.5), and (3.6) become fuzzy equations. For the uncertain nonlinear system (3.1), the following two types of fuzzy equations are used to model it

$$y_k = a_1 f_1(x_k) \oplus a_2 f_2(x_k) \oplus \cdots \oplus a_n f_n(x_k) \tag{3.26}$$

or

$$\begin{aligned} & a_1 f_1(x_k) \oplus a_2 f_2(x_k) \oplus \cdots \oplus a_n f_n(x_k) \\ & = b_1 g_1(x_k) \oplus b_2 g_2(x_k) \oplus \cdots \oplus b_m g_m(x_k) \oplus y_k. \end{aligned} \tag{3.27}$$

Because a_i and b_i are fuzzy numbers, the fuzzy operation $\oplus$ is used. Equation (3.27) has a more general form than (3.26); it is called a dual fuzzy equation.

In a special case, $f_i(x_k)$ has polynomial form,

$$y_k = a_1 x_k \oplus \cdots \oplus a_n x_k^n \tag{3.28}$$

or

$$a_1 x_k \oplus \cdots \oplus a_n x_k^n = b_1 x_k \oplus \cdots \oplus b_n x_k^n \oplus y_k. \tag{3.29}$$

Equation (3.28) is called a fuzzy polynomial and (3.29) is called a dual fuzzy polynomial.

Modeling with a fuzzy equation (or a fuzzy polynomial) can be considered as fuzzy interpolation. The polynomial fuzzy equation (3.28) or dual polynomial fuzzy equation (3.29) is used to model a nonlinear function

$$z_k = f(x_k). \tag{3.30}$$

The object is to minimize error between the two outputs y_k and z_k. Since y_k is a fuzzy number and z_k is a crisp number, the maximum of all points are used as the modeling error

$$\max_k |y_k - z_k| = \max_k |y_k - f(x_k)| = \max_k |\beta_k| \tag{3.31}$$

where $y_k = F(a(k), b(k), c(k))$, $\beta_k = F(\beta_1, \beta_2, \beta_3)$, which are defined in (3.9). From the definition of the absolute value of a triangular fuzzy number (3.22),

$$\max_k |\beta_k| = \max_k \left[\begin{array}{c} |a(k) - f(x_k)| \\ +|b(k) - f(x_k)| + |c(k) - f(x_k)| \end{array} \right]$$

$$\beta_1 = \max_k |a(k) - f(x_k)|$$

$$\beta_2 = \max_k \{b(k) + f(x_k)\}$$

$$\beta_3 = \max_k \{c(k) + f(x_k)\}. \tag{3.32}$$

The modeling problem (3.31) is to find $a(k), b(k)$, and $c(k)$, such that

$$\min_{a_k, b_k, c_k} \left\{ \max_k |\beta_k| \right\} = \min_{a_k, b_k, c_k} \left\{ \max_k |y_k - f(x_k)| \right\}. \tag{3.33}$$

From (3.32)

$$\beta_1 \geq |a(k) - f(x_k)|, \quad \beta_2 \geq b(k) + f(x_k), \quad \beta_3 \geq c(k) + f(x_k). \tag{3.34}$$

Equation (3.33) can be solved by the linear programming method for a fuzzy polynomial,

$$\begin{cases} \text{subject :} & \begin{array}{c} \min \beta_1 \\ \beta_1 + \sum_{j=0}^{n} a_j x_k^j \geq f(x_k) \\ \beta_1 - \sum_{j=0}^{n} a_j x_k^j \geq -f(x_k) \end{array} \end{cases} \tag{3.35}$$

$$\begin{cases} \text{subject :} & \begin{array}{c} \min \beta_2 \\ \beta_2 - \sum_{j=0}^{n} a_j x_k^j \geq f(x_k) \\ \beta_2 \geq 0 \end{array} \end{cases} \tag{3.36}$$

$$\begin{cases} \text{subject :} & \begin{array}{c} \min \beta_3 \\ \beta_3 - \sum_{j=0}^{n} \bar{a}_j x_k^j \geq f(x_k) \\ \beta_3 \geq 0 \end{array} \end{cases}. \tag{3.37}$$

Also, (3.33) can be solved by the linear programming method for a dual fuzzy polynomial,

$$\begin{cases} \text{subject :} & \begin{array}{c} \min \beta_1 \\ \beta_1 + \sum_{j=0}^{n} a_j x_k^j \ominus \sum_{j=0}^{n} b_j x_k^j \geq f(x_k) \\ \beta_1 - \left\{ \sum_{j=0}^{n} a_j x_k^j \ominus \sum_{j=0}^{n} b_j x_k^j \right\} \geq -f(x_k) \end{array} \end{cases} \tag{3.38}$$

$$\begin{cases} \quad \min \beta_2 \\ \text{subject :} \quad \beta_2 - \left[\sum_{j=0}^{n} \underline{a}_j x_k^j \ominus \sum_{j=0}^{n} \underline{b}_j x_k^j \right] \geq f(x_k) \\ \quad \beta_2 \geq 0 \end{cases} \tag{3.39}$$

$$\begin{cases} \quad \min \beta_3 \\ \text{subject :} \quad \beta_3 - \left[\sum_{j=0}^{n} \overline{a}_j x_k^j \ominus \sum_{j=0}^{n} \overline{b}_j x_k^j \right] \geq f(x_k) \\ \quad \beta_3 \geq 0 \end{cases} \tag{3.40}$$

where $\underline{a}_j$, $\underline{b}_j$, $\overline{a}_j$, and $\overline{b}_j$ are defined as in (3.12). In this way, the best approximation of $f(x_k)$ at point x_k is $y_k = F\left(a_k, b_k, c_k\right)$. The approximation error of β_k is minimized.

In this chapter, fuzzy Equation (3.26) and dual fuzzy Equation (3.27) are used to model the uncertain nonlinear system (3.1) such that the output of the plant y_k can follow the desired output y_k^*,

$$\min_{a_k} \left\| y_k - y_k^* \right\|. \tag{3.41}$$

This modeling object can be considered as: finding a_k for the following fuzzy equation

$$y_k^* = a_1 f_1(x_k) \oplus a_2 f_2(x_k) \oplus \cdots \oplus a_n f_n(x_k) \tag{3.42}$$

where $x_k = [y_{k-1}^T, y_{k-2}^T, \cdots u_k^T, u_{k-1}^T, \cdots]^T$.

The controller design process is to find u_k, such that the output of the plant y_k can follow desired output y_k^*, or the trajectory tracking error is minimized

$$\min_{u_k} \left\| y_k - y_k^* \right\|. \tag{3.43}$$

This control object can be considered as: finding a solution u_k for the following dual fuzzy equation

$$\begin{aligned} & a_1 f_1(x_k) \oplus a_2 f_2(x_k) \oplus \cdots \oplus a_n f_n(x_k) \\ & = b_1 g_1(x_k) \oplus b_2 g_2(x_k) \oplus \cdots \oplus b_m g_m(x_k) \oplus y_k^*. \end{aligned} \tag{3.44}$$

3.1.1 Fuzzy Parameter Estimation with Neural Networks

A neural network is designed to represent the fuzzy Equation (3.26), see Figure 3.1. The input to the neural network is $x(k)$, the output is the fuzzy number $\hat{y}_k$. The weight is a_i. The objective is to find a suitable weight a_i such that the output of the neural network $\hat{y}_k$ converges to the desired output y_k^*.

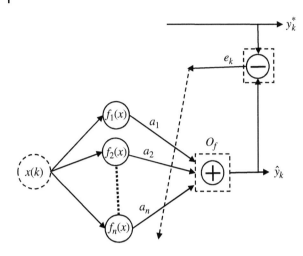

Figure 3.1 Fuzzy equation in the form of a neural network.

Figure 3.1 Fuzzy equation in the form of a neural network.

In order to simplify the operation of the neural network, the triangular fuzzy number (3.9) is used. The input fuzzy number x_k is first applied to the α-level as in (3.11)

$$[x_k]^\alpha = A\left(\underline{x_k}^\alpha, \overline{x_k}^\alpha\right) \quad k = 0 \cdots N. \tag{3.45}$$

Then

$$[O_j]^\alpha = A\left(f_j(\underline{x_k})^\alpha, f_j(\overline{x_k})^\alpha\right) \quad j = 1 \cdots n. \tag{3.46}$$

The output of the neural network is

$$[\hat{y}_k]^\alpha = A\left\{ \sum_{j\in M} \underline{O_j}^\alpha \underline{a_j}^\alpha + \sum_{j\in C} \overline{O_j}^\alpha \underline{a_j}^\alpha \right.$$
$$\left. + \underline{a_0}^\alpha, \sum_{j\in M'} \overline{O_j}^\alpha \overline{a_j}^\alpha + \sum_{j\in C'} \underline{O_j}^\alpha \overline{a_j}^\alpha + \overline{a_0}^\alpha \right\} \tag{3.47}$$

where $M = \{j | \underline{a_j}^\alpha \geq 0\}$, $C = \{j | \underline{a_j}^\alpha < 0\}$, $M' = \{j | \overline{a_j}^\alpha \geq 0\}$, $C' = \{j | \overline{a_j}^\alpha < 0\}$.

In order to train the weights, a cost function should be defined for the fuzzy numbers. The training error is

$$e_k = y_k^* - \hat{y}_k \tag{3.48}$$

where $[y_k^*]^\alpha = A\left(\underline{y_k^*}^\alpha, \overline{y_k^*}^\alpha\right)$, $[\hat{y}_k]^\alpha = A\left(\underline{\hat{y}_k}^\alpha, \overline{\hat{y}_k}^\alpha\right)$, $[e_k]^\alpha = A\left(\underline{e_k}^\alpha, \overline{e_k}^\alpha\right)$. The cost function is defined as

$$J_k = \underline{J}^\alpha + \overline{J}^\alpha$$

$$\underline{J}^\alpha = \frac{1}{2}\left(\underline{y_k^{*\alpha}} - \underline{\hat{y}_k}^\alpha\right)^2$$

$$\overline{J}^\alpha = \frac{1}{2}\left(\overline{y_k^{*}}^\alpha - \overline{\hat{y}_k}^\alpha\right)^2. \tag{3.49}$$

Obviously, $J_k \rightarrow 0$ means $\left[\hat{y}_k\right]^\alpha \rightarrow \left[y_k^*\right]^\alpha$.

Now the gradient method is used to train the weight $a_j = \left[a_j^1, a_j^2, a_j^3\right]$ defined in (3.9), or a_j^r, $r = 1, 2, 3$. $\frac{\partial J_k}{\partial a_j^r}$ is calculated as

$$\frac{\partial J_k}{\partial a_j^r} = \frac{\partial \underline{J}^\alpha}{\partial a_j^r} + \frac{\partial \overline{J}^\alpha}{\partial a_j^r}. \tag{3.50}$$

By the chain rule

$$\frac{\partial \underline{J}^\alpha}{\partial a_j^r} = \frac{\partial \underline{J}^\alpha}{\partial \underline{\hat{y}_k}^\alpha}\frac{\partial \underline{\hat{y}_k}^\alpha}{\partial a_j^\alpha}\frac{\partial a_j^\alpha}{\partial a_j^r} = \left(\underline{y_k^{*\alpha}} - \underline{\hat{y}_k}^\alpha\right)\Gamma \tag{3.51}$$

where $j = 0$, and

$$\frac{\partial \underline{J}^\alpha}{\partial a_j^r} = \frac{\partial \underline{J}^\alpha}{\partial \underline{\hat{y}_k}^\alpha}\frac{\partial \underline{\hat{y}_k}^\alpha}{\partial a_j^\alpha}\frac{\partial a_j^\alpha}{\partial a_j^r}$$

$$= \left(\underline{y_k^{*\alpha}} - \underline{\hat{y}_k}^\alpha\right)\begin{cases} \underline{O_j}^\alpha\Gamma, & a_j^\alpha \geq 0 \\ \overline{O_j}^\alpha\Gamma, & a_j^\alpha < 0 \end{cases} \tag{3.52}$$

where $j = 1, \ldots, n$ and $\Gamma = \begin{cases} 1-\alpha & r = 1 \\ \alpha & r = 2 \\ 0 & r = 3 \end{cases}$, also

$$\frac{\partial \overline{J}^\alpha}{\partial a_j^r} = \frac{\partial \overline{J}^\alpha}{\partial \overline{\hat{y}_k}^\alpha}\frac{\partial \overline{\hat{y}_k}^\alpha}{\partial \overline{a_j}^\alpha}\frac{\partial \overline{a_j}^\alpha}{\partial a_j^r} = \left(\overline{y_k^{*}}^\alpha - \overline{\hat{y}_k}^\alpha\right)\Gamma_1 \tag{3.53}$$

where $j = 0$, and

$$\frac{\partial \overline{J}^\alpha}{\partial a_j^r} = \frac{\partial \overline{J}^\alpha}{\partial \overline{\hat{y}_k}^\alpha}\frac{\partial \overline{\hat{y}_k}^\alpha}{\partial \overline{a_j}^\alpha}\frac{\partial \overline{a_j}^\alpha}{\partial a_j^r}$$

$$= \left(\overline{y_k^{*}}^\alpha - \overline{\hat{y}_k}^\alpha\right)\begin{cases} \overline{O_j}^\alpha\Gamma_1, & \overline{a_j}^\alpha \geq 0 \\ \underline{O_j}^\alpha\Gamma_1, & \overline{a_j}^\alpha < 0 \end{cases} \tag{3.54}$$

where $j = 1, \ldots, n$ and $\Gamma_1 = \begin{cases} 0 & r = 1 \\ \alpha & r = 2 \\ 1-\alpha & r = 3 \end{cases}$.

The coefficient a_j is updated as

$$a_j^r(k+1) = a_j^r(k) - \eta \frac{\partial J_k}{\partial a_j^r}$$

(3.55)

where $r = 1, 2, 3$ and η is the training rate $\eta > 0$. In order to increase training process, a momentum term is added as

$$a_j^r(k+1) = a_j^r(k) - \eta \frac{\partial J_k}{\partial a_j^r} + \gamma \left[a_j^r(k) - a_j^r(k-1) \right]$$

(3.56)

where $\gamma > 0$.

3.1.2 Upper Bounds of the Modeling Errors

In this section, some well known approximation theories are extended into fuzzy equation modeling. Initially, the modeling error is defined in the sense of a fuzzy number.

Definition 3.9 *The distance between two fuzzy numbers, $u, v \in E$, is defined as the Hausdorff metric $d_H(u, v)$,*

$$d_H(u, v) = \max\{\sup_{x \in u} \inf_{y \in v} |x - y|, \sup_{y \in v} \inf_{x \in u} |x - y|\}.$$

(3.57)

Lemma 3.1 *If $\theta \subset E$ is a compact set, then θ is uniformly support bounded, i.e. there is a compact set $U \subset R$, such that $\forall u \in \theta$,*

$$\mathrm{supp}(u) \subset U.$$

(3.58)

Lemma 3.2 *Let $u, v \in E$, and $\alpha \in (0, 1]$, $\lambda \in (0, +\infty]$, then: (i) if $f : R \to R$ is continuous, $[f(u)]^\alpha = f([u^\alpha])$ holds; (ii) if $f : R \to R$ is continuous, then $f(\mathrm{supp}(u)) = \mathrm{supp}(f(u))$.*

Proof: It is only sufficient to prove (ii) since (i) is obtained from [193]. Initially, it should be demonstrated that $\overline{f(A)} = f(\overline{A})$ for $A \subset R$. In fact, since $f(A) \subset f(\overline{A})$, and $f(\overline{A})$ is closed by the continuity of f, hence $\overline{f(A)} \subset f(\overline{A})$. On the other hand, for arbitrarily given $y \in f(\overline{A})$, there is a sequence $\{x_n | n \in N\} \subset R$, and a $x \in R$, such that $x_n \to x$ $(n \to +\infty)$, $y = f(x)$. The continuity of f implies $\lim_{n \to +\infty} f(x_n) = f(x) = y$. But $f(x_n) \in f(A)$, so $y \in \overline{f(A)}$. Hence $f(\overline{A}) \subset \overline{f(A)}$. Thus $\overline{f(A)} = f(\overline{A})$.

Considering

$$\mathrm{supp}(f(u)) = \overline{\{y \in R | f(u)(y) > 0\}}$$

$$f(\mathrm{supp}(u)) = f\left(\overline{\{x \in R | u(x) > 0\}} \right).$$

(3.59)

So

$$f(\text{supp}(u)) = \overline{f(\{x \in R | u(x) > 0\})} = \overline{\{f(x) \in R | u(x) > 0\}} \qquad (3.60)$$

holds. Since it may be easily proved that $\{y \in R | f(u)(y) > 0\} = \{f(x) | u(x) > 0\}$, therefore,

$$\text{supp}(f(u)) = f(\text{supp}(u)) \qquad (3.61)$$

which implies the lemma. ∎

Lemma 3.3 *Let $B \subset R$ be a compact set, and f, g be continuous on B, $h > 0$, moreover*

$$\forall x \in B, |f(x) - g(x)| < h. \qquad (3.62)$$

Then for each compact set $B_1 \subset B$, $|\sup_{x \in B_1} f(x) - \sup_{x \in B_1} g(x)| < h$.

Proof: Because of the facts that B_1 is a compact set and f, g are continuous on B_1, then there are $x_0 \in B_1$, $y_0 \in B_1$, such that

$$f(x_0) = \sup_{x \in B_1} f(x), \quad g(y_0) = \sup_{x \in B_1} g(x). \qquad (3.63)$$

Supposing $|f(x_0) - g(y_0)| \geq h$, so

$$f(x_0) - g(y_0) \leq -h, \quad \text{or} \quad f(x_0) - g(y_0) \geq h. \qquad (3.64)$$

In the first case (3.64), because $f(y_0) \leq f(x_0)$,

$$f(y_0) - g(y_0) \leq f(x_0) - g(y_0) \leq -h \Rightarrow |f(y_0) - g(y_0)| \geq h \qquad (3.65)$$

holds, which contradicts (3.62). In the second case (3.64), since $g(x_0) \leq g(y_0)$, the following is obtained

$$f(x_0) - g(x_0) \geq f(x_0) - g(y_0) \geq h \Rightarrow |f(x_0) - g(x_0)| \geq h \qquad (3.66)$$

which also contradicts (3.83). Therefore, (3.62) is not true, hence $-h < f(x_0) - g(y_0) < h$, so $|f(x_0) - g(x_0)| < h$, i.e. $|\sup_{x \in B_1} f(x) - \sup_{x \in B_1} g(x)| < h$. The proof is completed. ∎

Theorem 3.1 *Let $f : R \to R$ be a continuous function, then for each compact set $\theta \subset E_0$ (the set of all the bounded fuzzy set), and $\psi > 0$, there are $n \in N$, and $a_0, a_i \in E_0$, $i = 1, 2, \ldots, n$, such that*

$$\forall x \in \theta \quad \text{and} \quad \forall \tilde{x} \in R, \quad d(f(\tilde{x}), \sum_{i=1}^{n} f_i(x)a_i + a_0) < \psi \qquad (3.67)$$

where ψ is a finite number.

Proof: The proof of Theorem can be followed by the results below. ∎

If the function $f : R \to R, f$ can be extended by the extension principle to the fuzzy function, which is also written as $f : E_0 \to E$ as follows:

$$\forall u \in E_0, \quad f(u)(y) = \bigvee_{f(x)=y} \{u(x)\} \quad y \in R \tag{3.68}$$

f is called the extended function. Moreover, $cc(R)$ stands for the set of bounded closed intervals of R. Obviously

$$u \in E_0 \Rightarrow \forall \alpha \in (0,1], \quad [u]^\alpha \in cc(R). \tag{3.69}$$

Moreover

$$\text{supp}(u) \in cc(R). \tag{3.70}$$

So from now, suppose

$$\text{supp}(u) = [s_1(u), s_2(u)]. \tag{3.71}$$

Theorem 3.2 *Let $f : R \to R$ be a continuous function, then for each compact set $\theta \subset E_0$, $\varrho > 0$ and arbitrary $\varepsilon > 0$, there are $n \in N$, and $a_0, a_i \in E_0$, $i = 1, 2, \ldots, n$, such that*

$$\forall x \in \theta, \quad d(f(x), \sum_{i=1}^{n} f_i(x)a_i + a_0) < \varrho \tag{3.72}$$

where ϱ is a finite number. The lower and upper limits of the α-level set of the fuzzy function minimize to ϱ, but the center approaches ε.

Proof: Because $\theta \subset E_0$ is a compact set, hence by Lemma 3.1, let $U \subset R$ be the compact set corresponding to θ. $\forall \varepsilon > 0$, by the conclusions in [61], there are $n \in N$, and $a_0, a_i \in R, i = 1, 2, \ldots, n$, such that

$$\forall x \in U, \quad |f(x) - \sum_{i=1}^{n} f_i(x)a_i + a_0| < \varepsilon \tag{3.73}$$

holds. Let $g(x) = \sum_{i=1}^{n} f_i(x)a_i + a_0, x \in R$, then

$$\forall x \in U, \quad |f(x) - g(x)| < \varepsilon. \tag{3.74}$$

By Theorem 3.5, (3.72) holds. ∎

Theorem 3.3 *Supposing $\theta \subset E_0$ is compact, U the corresponding compact set of θ, and $f, g : R \to R$ are the continuous functions which satisfy the condition that, for given $h > 0$,*

$$\forall x \in U, \quad |f(x) - g(x)| < h \tag{3.75}$$

holds. Then $\forall u \in \theta, \quad d(f(u) - g(u)) \leq h$.

Proof: Let $u \in E$ and $\alpha \in (0,1]$. Because f, g are continuous, hence $[f(u)]^\alpha = f([u^\alpha])$ and $[g(u)]^\alpha = g([u^\alpha])$ hold by Lemma 3.2. Therefore, the following facts are obtained by the conclusions from [159],

$$d_H([f(u)]^\alpha - [g(u)]^\alpha) = d_H(f([u^\alpha]) - g([u^\alpha]))$$
$$= \sup_{|p|=1} \{|s(p, f([u^\alpha])) - s(p, g([u^\alpha]))|\} \tag{3.76}$$

Because for $p \in R: |p| = 1$, so

$$|s(p, f([u^\alpha]) - s(p, g([u^\alpha]))|$$
$$= |\sup\{py|y \in f([u^\alpha])\} - \sup\{py|y \in g([u^\alpha])\}|$$
$$= |\sup\{pf(x)|x \in [u]^\alpha\} - \sup\{pg(x)|x \in [u]^\alpha\}| \tag{3.77}$$

holds. And considering the conditions in the theorem, the following is obtained

$$\forall x \in [u]^\alpha, \quad |pf(x) - pg(x)| = |f(x) - g(x)| < h. \tag{3.78}$$

Therefore, by (3.75), (3.77), and Lemma 3.3, the following

$$\forall \alpha \in (0,1], d_H([f(u)]^\alpha, [g(u)]^\alpha) < h \Rightarrow d(f(u), g(u))$$
$$= \sup_{\alpha \in (0,1]} \{d_H([f(u)]^\alpha, [g(u)]^\alpha)\} \leq h \tag{3.79}$$

holds, which proves the theorem.

Arbitrary given $u \in E$ and $\alpha \in (0,1]$. Because f, g are continuous, hence $[f(u)]^\alpha = f([u^\alpha])$, $[g(u)]^\alpha = g([u^\alpha])$ holds by Lemma 3.2. Therefore, the following facts are obtained by the conclusions from [159],

$$d([f(u)]^\alpha - [g(u)]^\alpha) = d_H(f([u^\alpha]) - g([u^\alpha]))$$
$$= \sup_{|p|=1} \{|s(p, f([u^\alpha])) - s(p, g([u^\alpha]))|\}. \tag{3.80}$$

Because for $p \in R: |p| = 1$, so

$$|s(p, f([u^\alpha]) - s(p, g([u^\alpha]))| = |\sup\{py|y \in f([u^\alpha])\} - \sup\{py|y \in g([u^\alpha])\}|$$
$$= |\sup\{pf(x)|x \in [u]^\alpha\} - \sup\{pg(x)|x \in [u]^\alpha\}| \tag{3.81}$$

holds. Considering the conditions in the theorem, the following is obtained

$$\forall x \in [u]^\alpha, \quad |pf(x) - pg(x)| = |f(x) - g(x)| < h. \tag{3.82}$$

Therefore, by (3.75), (3.77) and Lemma 3.3, the following holds

$$\forall \alpha \in (0,1], \quad d_H([f(u)]^\alpha, [g(u)]^\alpha) < h \Rightarrow d(f(u), g(u))$$
$$= \sup_{\alpha \in (0,1]} \{d_H([f(u)]^\alpha, [g(u)]^\alpha)\} \leq h \tag{3.83}$$

which proves the theorem.

3.2 Control with Fuzzy Equations

There is no analytical solution for the dual fuzzy Equation (3.44). Here, neural networks are used to approximate the solution (control). In order to use neural networks to approximate the solution of the dual fuzzy equation (3.44), it is essential to transform it into normal fuzzy equation as (3.26).

Generally, the inverse element for an arbitrary fuzzy number $u \in E$ does not exist; there is no $v \in E$, such that

$$u \oplus v = 0. \tag{3.84}$$

In other words,

$$u \oplus (\ominus u) \neq 0. \tag{3.85}$$

So (3.44) cannot be

$$a_1 f_1(x_k) \oplus \cdots \oplus a_n f_n(x_k) \ominus b_1 g_1(x_k) \ominus \cdots \ominus b_m g_m(x_k) = y_k^*$$
$$\left[a_1 \ominus b_1\right] f_1(x_k) \oplus \left[a_2 \ominus b_2\right] f_2(x_k) \oplus \cdots = y_k^*. \tag{3.86}$$

Here the $\varnothing$ operation is used. $\oplus b_i g_i(x)$ is added and $\varnothing \tau$ is applied on both sides of (3.44)

$$a_1 f_1(x_k) \oplus \cdots \oplus a_n f_n(x_k) \oplus \left\{ \left[b_1 g_1(x) \oplus \cdots \oplus b_m g_m(x)\right] \varnothing \tau \right\}$$
$$= b_1 g_1(x_k) \oplus \cdots \oplus b_m g_m(x_k) \oplus \left\{ \left[b_1 g_1(x) \oplus \cdots \oplus b_m g_m(x)\right] \varnothing \tau \right\} \oplus y_k^*. \tag{3.87}$$

When $\tau = 1$, by the definition of $\varnothing$, (3.87) is

$$a_1 f_1(x) \oplus \cdots \oplus a_n f_n(x) \ominus b_1 g_1(x) \ominus \cdots \ominus b_m g_m(x) = y_k^*. \tag{3.88}$$

A neural network is designed to represent the fuzzy equation (3.88), see Figure 3.2. The input to the neural network is the fuzzy numbers a_i and b_i; the output of the fuzzy number is y_k. The weights are $f_i(x)$ and $g_j(x)$,

The objective is to find a suitable weight x (solution) such that the output of the neural network $\hat{y}_k$ converges to the desired output y_k^*. From the control point of view, it is essential to find a controller u_k that is a function of x such that the output of the plant (3.1) y_k (crisp value) approximates the fuzzy number y_k^*.

In order to simplify the operation of the neural network as in Figure 3.2, the triangular fuzzy number (3.9) is used. The input fuzzy numbers a_i and b_i are initially applied to the α-level as in (3.11)

$$[a_i]^\alpha = A\left(\underline{a}_i^\alpha, \overline{a}_i^\alpha\right) \quad i = 1 \cdots n$$
$$[b_j]^\alpha = A\left(\underline{b}_i^\alpha, \overline{b}_i^\alpha\right) \quad j = 1 \cdots m. \tag{3.89}$$

Figure 3.2 Dual fuzzy
equation in the form of a
neural network (NN).

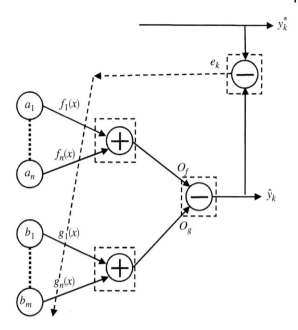

Then they are multiplied by the weights $f_i(x)$ and $g_j(x)$, and summarized according to (3.14)

$$[O_f]^\alpha = A \begin{pmatrix} \sum_{i \in M_f} f_i(x) \underline{a_i}^\alpha + \sum_{i \in C_f} f_i(x) \overline{a_i}^\alpha, \\ \sum_{i \in C_f} f_i(x) \overline{a_i}^\alpha, \sum_{i \in M_f} f_i(x) \underline{a_i}^\alpha \end{pmatrix}$$

$$[O_g]^\alpha = A \begin{pmatrix} \sum_{j \in M_g} g_j(x) \underline{b_j}^\alpha + \sum_{j \in C_g} g_j(x) \overline{b_j}^\alpha, \\ \sum_{j \in C_g} g_j(x) \overline{b_j}^\alpha, \sum_{j \in M_g} g_j(x) \underline{b_j}^\alpha \end{pmatrix} \qquad (3.90)$$

where $M_f = \{i | f_i(x) \geq 0\}$, $C_f = \{i | f_i(x) < 0\}$, $M_g = \{j | g_j(x) \geq 0\}$, and $C_g = \{j | g_j(x) < 0\}$.

The output of the neural network is

$$[\hat{y}_k]^\alpha = A \left(\underline{O_f}^\alpha - \underline{O_g}^\alpha, \overline{O_f}^\alpha - \overline{O_g}^\alpha \right). \qquad (3.91)$$

In order to train the weights, a cost function should be defined for the fuzzy numbers. The training error is

$$e_k = y_k^* \ominus \hat{y}_k \qquad (3.92)$$

where $\left[y_k^*\right]^\alpha = A\left(\underline{y}_k^{*\alpha}, \overline{y}_k^{*\alpha}\right)$, $\left[\hat{y}_k\right]^\alpha = A\left(\underline{\hat{y}}_k^\alpha, \overline{\hat{y}}_k^\alpha\right)$, and $\left[e_k\right]^\alpha = A\left(\underline{e}_k^\alpha, \overline{e}_k^\alpha\right)$. The cost function is defined as

$$J_k = \underline{J}^\alpha + \overline{J}^\alpha$$

$$\underline{J}^\alpha = \frac{1}{2}\left(\underline{y}_k^{*\alpha} - \underline{\hat{y}}_k^\alpha\right)^2$$

$$\overline{J}^\alpha = \frac{1}{2}\left(\overline{y}_k^{*\alpha} - \overline{\hat{y}}_k^\alpha\right)^2. \tag{3.93}$$

Obviously, $J_k \to 0$ means $\left[\hat{y}_k\right]^\alpha \to \left[y_k^*\right]^\alpha$.

Remark 3.1 *The main advantage of the least mean square index (3.93) is that it has a self-correcting feature that permits operation for an arbitrarily long period without deviating from its constraints. The corresponding gradient algorithm is susceptible to cumulative round-off errors and is suitable for long runs without an additional error correction procedure. It is more robust in statistics, identification, and signal processing* [180].

Now the gradient method is used to train the weights $f_i(x)$ and $g_j(x)$. The solution x_0 is the functions of $f_i(x)$ and $g_j(x)$. $\frac{\partial J_k}{\partial x_0}$ is calculated as

$$\frac{\partial J_k}{\partial x_0} = \frac{\partial \underline{J}^\alpha}{\partial x_0} + \frac{\partial \overline{J}^\alpha}{\partial x_0}. \tag{3.94}$$

By the chain rule

$$\frac{\partial \underline{J}^\alpha}{\partial x_0} = \frac{\partial \underline{J}^\alpha}{\partial \underline{\hat{y}}_k^\alpha} \frac{\partial \underline{\hat{y}}_k^\alpha}{\partial \underline{O}_f^\alpha} \sum \frac{\partial \underline{O}_f^\alpha}{\partial f_i(x)} \frac{\partial f_i(x)}{\partial x_0} + \frac{\partial \underline{e}^\alpha}{\partial \underline{\hat{y}}_k^\alpha} \frac{\partial \underline{\hat{y}}_k^\alpha}{\partial \underline{O}_g^\alpha} \sum \frac{\partial \underline{O}_g^\alpha}{\partial g_j(x)} \frac{\partial g_j(x)}{\partial x_0}$$

$$\frac{\partial \overline{J}^\alpha}{\partial x_0} = \frac{\partial \overline{J}^\alpha}{\partial \overline{\hat{y}}_k^\alpha} \frac{\partial \overline{\hat{y}}_k^\alpha}{\partial \overline{O}_f^\alpha} \sum \frac{\partial \overline{O}_f^\alpha}{\partial f_i(x)} \frac{\partial f_i(x)}{\partial x_0} + \frac{\partial \overline{e}_k^\alpha}{\partial \overline{\hat{y}}_k^\alpha} \frac{\partial \overline{\hat{y}}_k^\alpha}{\partial \overline{O}_f^\alpha} \sum \frac{\partial \overline{O}_f^\alpha}{\partial g_j(x)} \frac{\partial g_j(x)}{\partial x_0}. \tag{3.95}$$

If f_i' and g_j' are positive

$$\frac{\partial \underline{J}^\alpha}{\partial x_0} = \sum_{i=1}^n -\left(\underline{y}_k^{*\alpha} - \underline{\hat{y}}_k^\alpha\right) \underline{a}_i^\alpha f_i' + \sum_{j=1}^m \left(\underline{y}_k^{*\alpha} - \underline{\hat{y}}_k^\alpha\right) \underline{b}_j^\alpha g_j'$$

$$\frac{\partial \overline{J}^\alpha}{\partial x_0} = \sum_{i=1}^n -\left(\overline{y}_k^{*\alpha} - \overline{\hat{y}}_k^\alpha\right) \overline{a}_i^\alpha f_i' + \sum_{j=1}^m \left(\overline{y}_k^{*\alpha} - \overline{\hat{y}}_k^\alpha\right) \overline{b}_j^\alpha g_j'. \tag{3.96}$$

Otherwise

$$\frac{\partial \underline{J}^\alpha}{\partial x_0} = \sum_{i=1}^n -\left(\underline{y}_k^{*\alpha} - \underline{\hat{y}}_k^\alpha\right) \overline{a}_i^\alpha f_i' + \sum_{j=1}^m \left(\underline{y}_k^{*\alpha} - \underline{\hat{y}}_k^\alpha\right) \overline{b}_j^\alpha g_j'$$

Figure 3.3 Dual fuzzy equation in the form of a feedback neural network (FNN).

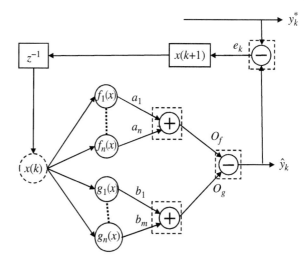

$$\frac{\partial \overline{J}^{-\alpha}}{\partial x_0} = \sum_{i=1}^{n} -\left(\overline{y}_k^{*-\alpha} - \overline{\hat{y}}_k^{-\alpha}\right) \underline{a}_i^\alpha f_i' + \sum_{j=1}^{m} \left(\overline{y}_k^{*-\alpha} - \overline{\hat{y}}_k^{-\alpha}\right) \underline{b}_j^\alpha g_j'. \tag{3.97}$$

The solution x_0 is updated as

$$x_0\,(k+1) = x_0\,(k) - \eta \frac{\partial J_k}{\partial x_0} \tag{3.98}$$

where η is the training rate $\eta > 0$. In order to increase the training process, a momentum term is added as

$$x_0\,(k+1) = x_0\,(k) - \eta \frac{\partial J_k}{\partial x_0} + \gamma \left[x_0\,(k) - x_0\,(k-1)\right] \tag{3.99}$$

where $\gamma > 0$.

After x_0 is updated, it should be substituted into the weights $f_i\left(x_0\right)$ and $g_j\left(x_0\right)$.

The solution of the dual fuzzy equation (3.44) can be also approximated by another type of neural network, see Figure 3.3. Here the inputs are the nonlinear functions $f_i\,(x)$ and $g_j\,(x)$, the weights are the fuzzy numbers a_i and b_j. The training error e_k is used to update x.

The input is a crisp number $x\,(k)$. The nonlinear operations $f_i\,(x)$ and $g_j\,(x)$, and O_f and O_g are the same as in (3.90). The output of this neural network is the same as for (3.91).

The difference between the networks of Figure 3.2 (NN) and Figure 3.3 (FNN) is that the FNN does not change weights, it is an autonomous system. An NN is a standard neural network. An FNN is more robust than NN, and a bigger training rate η can be used in the FNN.

Since the control object is to find a u_k for the dual fuzzy equation (3.44), the controllability problem occurs if the dual fuzzy equation has a solution.

In order to show the existence of the solution of (3.44), the following lemmas are required

Lemma 3.4 *If the dual fuzzy equation* (3.34) *has a crisp solution u_k, then*

$$\left\{ \cap_{j=1}^{n}\text{domain}\left[f_j\left(x\right)\right] \right\} \cap \left\{ \cap_{j=1}^{m}\text{domain}\left[g_j\left(x\right)\right] \right\} \neq \phi. \tag{3.100}$$

Proof: Let $u_0 \in R$ be a solution of (3.44), the dual fuzzy equation then becomes

$$a_1 f_1(u_0) \oplus \cdots \oplus a_n f_n(u_0) = b_1 g_1(u_0) \oplus \cdots \oplus b_m g_m(u_0) \oplus y_k^*. \tag{3.101}$$

Since $f_j(u_0)$ and $g_j(u_0)$ exist, $u_0 \in$ domain $\left[f_j\left(x\right)\right]$ and $u_0 \in$ domain $\left[g_j\left(x\right)\right]$. Consequently, it can be concluded that $u_0 \in \cap_{j=1}^{n}\text{domain}\left[f_j\left(x\right)\right] = D_1$ and $u_0 \in \cap_{j=1}^{m}\text{domain}\left[g_j\left(x\right)\right] = D_2$. So there exists u_0, such that $u_0 \in D_1 \cap D_2 \neq \phi$. ∎

Obviously, the necessary condition for the existence of the solution of (3.44) is (3.100).

Assume two fuzzy numbers m_0 and $n_0 \in E$, $m_0 < n_0$. Define a set $K(x) = \{x \in E, m_0 \leq x \leq n_0\}$, and an operator $S : K \to K$, such that

$$S\left(m_0\right) \geq m_0, \quad S\left(n_0\right) \leq n_0 \tag{3.102}$$

where S is condensing and continuous, is bounded as $S(z) < r(z)$, $z \subset K$, and $r(z) > 0$. $r(Z)$ can be regarded as the measure of z.

Lemma 3.5 *If $n_i = S\left(n_{i-1}\right)$ and $m_i = S\left(m_{i-1}\right), i = 1, 2, \ldots$, and the upper and lower bounds of S are $\bar{s}$ and $\underline{s}$, then*

$$\bar{s} = \lim_{i \to +\infty} n_i, \quad \underline{s} = \lim_{i \to +\infty} m_i, \tag{3.103}$$

and

$$m_0 \leq m_1 \leq \ldots \leq m_n \leq \ldots \leq n_n \leq \ldots \leq n_1 \leq n_0. \tag{3.104}$$

The proof of this lemma is straightforward, see [57].

If there exists a fixed point x_0 in K, the successive iterates $x_i = S\left(x_{i-1}\right)$, $i = 1, 2, \ldots$ will converge to x_0, i.e. the distance (3.21) $\lim_{i \to \infty} d(x_i, x_0) = 0$.

Theorem 3.4 *If the fuzzy numbers a_i and b_j $(i = 1 \ldots n, j = 1 \ldots m)$ in (3.44) satisfy the Lipschitz condition* (3.13)

$$\begin{aligned} \left|d_M(a_i) - d_M(a_k)\right| &\leq H \left|a_i - a_k\right| \\ \left|d_U(a_i) - d_U(a_k)\right| &\leq H \left|a_i - a_k\right| \\ \left|d_M(b_i) - d_M(b_k)\right| &\leq H \left|b_i - b_k\right| \\ \left|d_U(b_i) - d_U(b_k)\right| &\leq H \left|b_i - b_k\right| \end{aligned} \tag{3.105}$$

where $k = 1 \ldots n$, d_M and d_U are defined in (3.13), and the upper bounds of f_i and g_j are $|f_i| \leq \overline{f}$ and $|g_j| \leq \overline{g}$, then the dual fuzzy equation (3.44) has a solution u which is in the following set

$$K_H = \left\{ \begin{array}{c} u \in E, \left|\overline{u}^{\alpha_1} - \underline{u}^{\alpha_2}\right| \\ \leq \left(n\overline{f} + m\overline{g}\right) H \left|\alpha_1 - \alpha_2\right| \end{array} \right\}. \tag{3.106}$$

Proof: Because the fuzzy numbers a_i and b_j in (3.44) are linear-in-parameter, from the definition (3.13) and the property (3.18)

$$d_M(\alpha) = a_{1M}(\alpha)f_1(x) \oplus \cdots \oplus a_{nM}(\alpha)f_n(x) \ominus b_{1M}(\alpha)g_1(x) \ominus \cdots \ominus b_{mM}(\alpha)g_m(x). \tag{3.107}$$

So

$$\begin{aligned} \left|d_M(\alpha) - d_M(\varphi)\right| &= \left|f_1(x)\right| \left| a_{1M}(\alpha) \ominus a_{1M}(\varphi) \right| \\ &+ \cdots + \left|f_n(x)\right| \left|a_{nM}(\alpha) \ominus a_{nM}(\varphi)\right| \\ &+ \left|g_1(x)\right| \left|b_{1M}(\alpha) \ominus b_{1M}(\varphi)\right| \\ &+ \cdots + \left|g_m(x)\right| \left|b_{mM}(\alpha) \ominus b_{mM}(\varphi)\right|. \end{aligned} \tag{3.108}$$

By the Lipschitz condition (3.13), (3.108) is

$$\begin{aligned} \left|d_M(\alpha) - d_M(\varphi)\right| &\leq \overline{f}H \sum_{i=1}^{n} |\alpha - \varphi| + \overline{g}H \sum_{i=1}^{n} |\alpha - \varphi| \\ &= \left(n\overline{f} + m\overline{g}\right) H |\alpha - \varphi|. \end{aligned} \tag{3.109}$$

Similarly, the upper bounds satisfy

$$\left|d_U(\alpha) - d_U(\varphi)\right| \leq \left(n\overline{f} + m\overline{g}\right) H |\alpha - \varphi|. \tag{3.110}$$

Since the lower bound $\left|d_M(\alpha) - d_M(\varphi)\right| \geq 0$, by Lemma 3.2 the solution is in K_H, which is defined in (3.106). ∎

The following theorem uses linear programming conditions (3.38)–(3.40) to show the controllability conditions of the dual polynomial fuzzy equation (3.29).

Lemma 3.6 *If the data number m and the order of the polynomial n in (3.29) satisfy*

$$m \geq 2n + 1 \tag{3.111}$$

where $k = 1 \ldots m$, then the solutions of (3.39) and (3.40) are $\beta_2 = \beta_3 = 0$.

Proof: Because

$$\sum_{j=0}^{n} \underline{a}_j x_k^j \ominus \sum_{j=0}^{n} \underline{b}_j x_k^j \leq -f(x_k) \tag{3.112}$$

$i = 1, 2, \ldots, m$, $2n + 1$ points are chosen for x_k, and the interpolating the dual polynomial

$$b(k) = \sum_{j=0}^{n} \underline{a}_j x_k^j \ominus \sum_{j=0}^{n} \underline{b}_j x_k^j. \tag{3.113}$$

If $h = \max_k \{b(k) + f(x_k)\}$ and $h > 0$, then the dual polynomial (3.29) can be changed into a new dual polynomial $b(k) - h$. This new dual polynomial satisfies (3.112). Because the feasible point of (3.39) $\beta_2 \geq 0$, it must be zero. A similar result can be obtained for (3.40).

Both $f(x_k)$ and x_k are crisp. If the data number is $k = 1 \ldots n$, there exists a solution for the polynomial approximation [126]. Because $b(k)$ and $c(k)$, (3.38) has a solution. ■

Theorem 3.5 *If the data number is big enough as (3.111), and the dual polynomial fuzzy equation (3.29) satisfies*

$$D\left[h\left(x_{k1}, u_{k1}\right), h\left(x_{k2}, u_{k2}\right)\right] \leq lD\left[u_{k1}, u_{k2}\right] \tag{3.114}$$

where $0 < l < 1$, $h(\cdot)$ represents a dual polynomial fuzzy equation,

$$h\left(x_{k1}, u_{k1}\right) : a_1 x_{k1} \oplus \cdots \oplus a_n x_{k1}^n = b_1 x_{k1} \oplus \cdots \oplus b_n x_{k1}^n \oplus y_{k1}. \tag{3.115}$$

$D[u, v]$ is the Hausdorff distance [178],

$$D[u, v] = \max\left\{\sup_{x \in u} \inf_{y \in v} d(x, y), \sup_{x \in v} \inf_{y \in u} d(x, y)\right\} \tag{3.116}$$

$d(x, y)$ is the distance defined in (3.21), then (3.29) has a unique solution u.
Proof: From Lemma 3.2, there are solutions for (3.38)–(3.40), if there are many data that satisfy (3.111). Without loss of generality, it can be assumed that the solutions for (3.38)–(3.40) are at $x_k = 0$, which corresponds to u_0. Equation (3.114) means $h(\cdot)$ in (3.115) is continuous. If $\delta > 0$ is chosen in such a manner that $D\left[y_k, u_0\right] \leq \delta$, then

$$D\left[h(x_k, u_0), u_0\right] \leq (1 - l)\delta. \tag{3.117}$$

Here $h(0, u_0) = u_0$. Now x can be selected around 0, $x_k \in [0, c]$, $c > 0$, and define

$$C_0 : \rho = \sup_{x_k \in [0, c]} D\left[y_{k_1}, y_{k_2}\right]. \tag{3.118}$$

Let $\{y_{k_m}\}$ be a sequence in C_0; for any $\varepsilon > 0$, $N_0(\varepsilon)$ can be found such that $\rho < \varepsilon$, $m, n \geq N_0$. So $y_{k_m} \to y_k$ for $x_k \in [0, c]$. Furthermore

$$D\left[y_k, u_0\right] \leq D\left[y_k, y_{k_m}\right] + D\left[y_{k_m}, u_0\right] < \varepsilon + \delta \tag{3.119}$$

for all $x \in [0, c]$, $m \geq N_0(\varepsilon)$. Since $\varepsilon > 0$ is arbitrary small,

$$D\left[y_k, u_0\right] \leq \delta \tag{3.120}$$

for all $x \in [0, c]$. Now it should be shown that y_k is continuous at $x_0 = 0$. Given $\delta > 0$, there exists $\delta_1 > 0$ such that

$$D\left[y_k, u_0\right] \leq D\left[y_k, y_{k_m}\right] + D\left[y_{k_m}, u_0\right] \leq \varepsilon + \delta_1 \tag{3.121}$$

for every $m \geq N_0(\varepsilon)$, by (3.120), whenever $|x - x_0| < \delta_1$, y_k is continuous at $x_0 = 0$. So (3.29) has a unique solution u_0. ∎

The necessary condition for the controllability (existence of solution) of the dual fuzzy equation (3.44) is (3.100), the sufficient condition of the controllability is (3.105). For most of membership functions such as the triangular function (3.9) and the trapezoidal function (3.10), the Lipschitz condition (3.105) is satisfied. They are controllable.

3.3 Simulations

In this section, several real applications are used to show how to use the dual fuzzy equation to design a fuzzy controller.

Example 3.1 A chemistry process   In a chemical reaction the poly ethylene (PE) and poly propylene (PP) are used to generate a desired substance (DS). If the cost of the material is defined as x, the cost PE is x and the cost of PP is x^2. The weights of PE and PP are uncertain, which satisfy the triangle function (3.9). A chemical factory wants to produce two types of DS. If a chemical factory wishes the cost between them to be $F(3.5, 4, 5) = y^*$, what is the cost x ? The weights of PE are $F(2.5, 3, 3.25) = a_1$ and $F(0.75, 1, 1.25) = b_1$. The weights of PP are $F(1.75, 2, 2.5) = a_2$ and $(1.75, 2, 2.5) = b_2$. The above relation can be modeled by the following dual fuzzy equation

$$(2.5, 3, 3.25)x \oplus (1.75, 2, 2.5)x^2$$
$$= (0.75, 1, 1.25)x \oplus (1.75, 2, 2.5)x^2 \oplus (3.5, 4, 5). \tag{3.122}$$

Here $f_1(x) = g_1(x) = x$, $f_2(x) = g_2(x) = x^2$. The NN and FNN shown in Figures 3.2 and 3.3 are used to approximate the solution x. The learning rates for both neural networks are the same, $\eta = 0.02$. The results are shown in Table 3.1. The exact solution is $x_0 = 2$. The neural networks start from $x(0) = 4$. Both neural networks converge to the real solution. The error $|\hat{x} - x_0|$ between the approximate solution $\hat{x}$ and the exact solution x_0 is shown in Figure 3.4. ∎

Table 3.1 Comparison results of the chemistry process.

k	x (k) with NN	k	x (k) with FNN
1	3.8377	1	3.7970
2	3.6105	2	3.3090
3	3.3435	3	2.9567
⋮	⋮	⋮	⋮
38	2.0053	26	2.0080
39	2.0044	27	2.0053
40	2.0036	28	2.0034

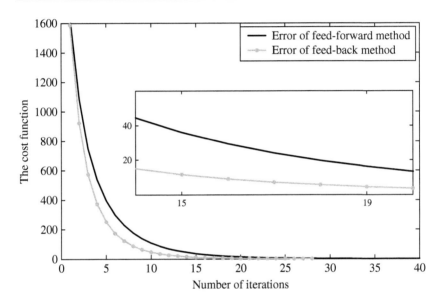

Figure 3.4 The error between the approximate solution and the exact solution.

Example 3.2 Heat source by insulating materials A heat source is in the center of insulating materials. The thickness of the materials is not exact, which satisfies the trapezoidal function (3.10),

$$A = F(0.12, 0.14, 0.15, 0.18) = a_1$$
$$B = F(0.08, 0.1, 0.2, 0.5) = a_2$$
$$C = F(0.09, 0.1, 0.2, 0.4) = b_1$$
$$D = F(0.02, 0.03, 0.05, 0.08) = b_2. \tag{3.123}$$

See Figure 3.5. The conductivity coefficients of these materials are $K_A = e^x = f_1$, $K_B = x\sqrt{x} = f_2$, $K_C = x^2 = g_1$, $K_D = x\sin\left(\frac{\Pi x}{8}\right) = g_2$, and x is the

Figure 3.5 Heat source by insulating materials.

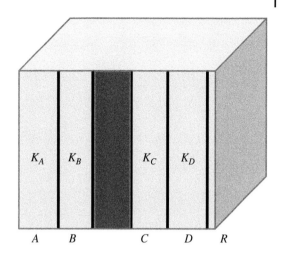

Table 3.2 Comparison results of the heat source.

k	x (k) with NN	k	x (k) with FNN
1	0.6251	1	0.7250
2	1.0542	2	1.1060
3	1.3321	3	1.5042
⋮	⋮	⋮	⋮
39	2.9899	10	2.9931
40	2.9922	11	2.9959
41	2.9940	12	2.9974

elapsed time. The object of the example is to find the time when the thermal resistance at the right side is $R = F(0.00415, 0.00428, 0.00569, 0.03187) = y^*$. The thermal balance is [81][130]:

$$\frac{A}{K_A} \oplus \frac{B}{K_B} = \frac{C}{K_C} \oplus \frac{D}{K_D} \oplus R. \tag{3.124}$$

The exact solution is $x = 3$ [81]. The maximum learning rate of the NN as Figure 3.2 is $\eta = 0.005$. The maximum learning rate of the FNN as Figure 3.3 is $\eta = 0.1$. The approximation results are shown in Table 3.2. The FNN is faster and more robust than the NN. ∎

Example 3.3 Water tank system The water tank system has two inlet valves q_1, q_2, and two outlet valves q_3, q_4, see Figure 1.1. The areas of the valves are uncertain as the triangle function (3.9), $A_1 = F(0.023, 0.025, 0.026)$, $A_2 = F(0.01, 0.02, 0.04)$, $A_3 = F(0.014, 0.015, 0.017)$, $A_4 = F(0.04, 0.06, 0.07)$.

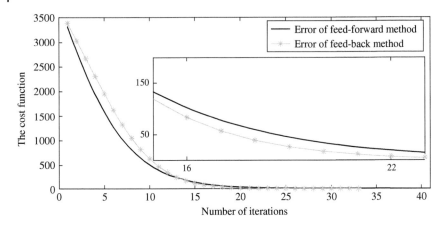

Figure 3.6 The error between the approximate solution and the exact solution.

The velocities of the flow (controlled by the valves) are $f_1 = (\frac{x}{10})e^x$, $f_2 = x\cos(\Pi x)$, $f_3 = \cos\left(\frac{\Pi x}{8}\right)$, $f_4 = \frac{x}{2}$. If it is a requirement that the outlet flow be $q = (4.090, 6.338, 36.402) = y^*$, what is the control variable x. The mass balance of the tank is [177]:

$$\rho A_1 f_1 \oplus \rho A_2 f_2 = \rho A_3 f_3 \oplus \rho A_4 f_4 \oplus q \qquad (3.125)$$

where ρ is the density of the water. The exact solution is $x_0 = 2$ [177]. It has been supposed that $x(0) = 5, \eta = 0.001, \gamma = 0.001$ for both the NN and the FNN. The error $|\hat{x} - x_0|$ between the approximate solution $\hat{x}$ and the exact solution x_0 is shown in Figure 3.6. For this example, both the NN and FNN work well. ∎

Example 3.4 Solid cylindrical rod The deformation of a solid cylindrical rod depends on the stiffness E, the forces on it F, the positions of the forces L, and the diameter of the rod d [181], see Figure 1.3. The positions are not exact, they satisfy the trapezoidal function (3.10). $L_1 = F(0.3, 0.4, 0.6, 0.7)$, $L_2 = F(0.5, 0.7, 0.8, 0.9)$, and $L_3 = F(0.5, 0.7, 0.8, 0.9)$. The area of the rod is $A = \frac{\pi}{4}d^2$. The external forces are the function of x, $F_1 = x^7$, $F_2 = x^6\sqrt{x}$, $F_3 = e^{2x}$ [42]. It is the requirement that the desired deformation at the point N be $N^* = F(0.000673, 0.000931, 0.001164, 0.001310)$ as in (3.10). What is the amount of control force that should be applied? According to the tension relations [42]

$$\frac{L_1 F_1}{AE} \oplus \frac{L_2(F_1 + F_2)}{AE} = \frac{L_3 F_3}{AE} \oplus N^* \qquad (3.126)$$

where $d = 0.02, E = 70 \times 10^9$. The exact solution is $x = 4$.

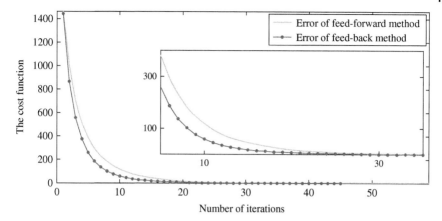

Figure 3.7 The error between the approximate solution and the exact solution.

Figure 3.8 Water channel system.

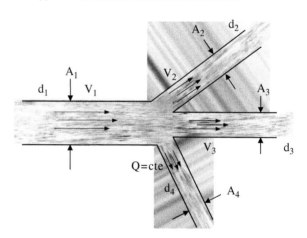

It has been supposed that $x(0) = 7$, $\eta = 0.002$, and $\gamma = 0.002$ for both the NN and FNN. The error $|\hat{x} - x_0|$ between the approximate solution $\hat{x}$ and the exact solution x_0 is shown in Figure 3.7. For this example, both the NN and FNN work well. The FNN is a little better than the NN. ∎

Example 3.5 Water Channel system The water in the pipe d_1 is divided into three pipes d_2, d_3, and d_4, see Figure 3.8. The areas of the pipes are uncertain, they satisfy the trapezoidal function (3.10). $A_1 = F(0.4, 0.6, 0.7, 0.8)$, $A_2 = F(0.05, 0.1, 0.2, 0.4)$, and $A_3 = F(0.03, 0.08, 0.1, 0.2)$. The water velocities in the pipes are controlled by the valve parameters x, $v_1 = x^3$, $v_2 = \frac{e^x}{2}$, and $v_3 = x$ [177]. The control object is to let the flow in pipe d_4, which is

$$Q = F(10.207861, 14.955723, 16.591446, 16.982892). \tag{3.127}$$

Table 3.3 Comparison results of the water channel system.

k	x (k) with NN	k	x (k) with FNN
1	5.9024	1	5.9226
2	5.7361	2	5.5341
3	5.5321	3	5.1234
⋮	⋮	⋮	⋮
77	3.0599	21	3.0162
78	3.0322	22	3.0131
79	3.0110	23	3.0086

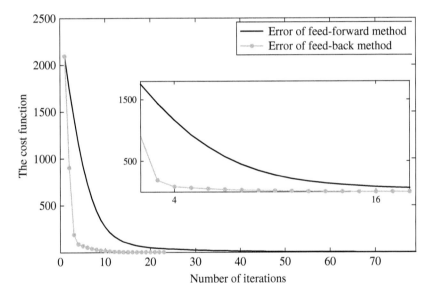

Figure 3.9 The error between the approximate solution and the exact solution.

What is the valve control parameter x? By mass balance

$$A_1 v_1 = A_2 v_2 \oplus A_3 v_3 \oplus Q. \tag{3.128}$$

The exact solution is $x = 3$ [177]. The maximum learning rate of the NN as Figure 3.2 is $\eta = 0.001$. The maximum learning rate of the FNN as Figure 3.3 is $\eta = 0.08$. The approximation results are shown in Table 3.3. The error $|\hat{x} - x_0|$ between the approximate solution $\hat{x}$ and the exact solution x_0 is shown in Figure 3.9. The FNN is faster and more robust than the NN. ∎

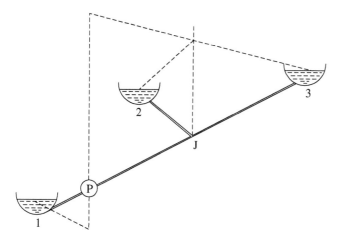

Figure 3.10 Pumping water from one tank to the other two tanks.

Example 3.6 Water pumping system There are three tanks connected to a pipeline at a constant H, see Figure 3.10. It is required to pump water from one tank to the other two tanks. This system satisfies the following relation

$$H = A_0 \oplus A_1 Q \oplus A_2 Q^2 \oplus A_3 Q^3 \tag{3.129}$$

where Q is the quantity of flow, H is the height of the pipe, A_0, A_1, A_2, and A_3 are the characteristic coefficients of the pump., they are

$$A_0 = (1, 5, 8), A_1 = (3, 7, 8), A_2 = (1, 2, 4), A_3 = (1, 3, 4) \tag{3.130}$$

The following 4 real uncertain data are taken into consideration

$$Q = \{2, (2, 4, 5), (3, 5, 6, 7), (1, 2, 4)\} \tag{3.131}$$

where $(2, 4, 5)$ and $(1, 2, 4)$ satisfy the triangle function (3.9), $(3, 5, 6, 7)$ is the trapezoidal function (3.10), 2 is a crisp number.

$$H = \{(19, 51, 72), (19{,}257, 648), (46{,}465, 767, 1632), (6, 51{,}360)\} . \tag{3.132}$$

Now a neural network is used to approximate A_0, A_1, A_2, and A_3. The results are shown in Table 3.4 and Table 3.5. ∎

Example 3.7 Heat source by insulating materials The heat source is in the left of the insulating materials, see Figure 3.11. The conductivity coefficients of these materials are $K_A = e^x = f_1$, $K_B = x\sqrt{x} = f_2$, $K_C = x^2 = g_1$, $K_D = x \sin\left(\frac{\Pi x}{8}\right) = g_2$, where x is the elapsed time. The thermal balance is [81]:

$$R = \frac{A}{K_A} \oplus \frac{B}{K_B} \oplus \frac{C}{K_C} \oplus \frac{D}{K_D}. \tag{3.133}$$

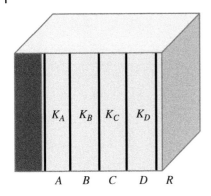

Figure 3.11 Heat source by insulating materials.

Table 3.4 Neural network approximation for the coefficients.

k	$\hat{A}_0$	$\hat{A}_1$
1	(3.9, 7.9, 10.9)	(5.9, 9.8, 10.9)
2	(3.7, 7.7, 10.6)	(5.7, 9.6, 10.7)
⋮	⋮	⋮
75	(1.0, 5.0, 8.0)	(3.0, 7.0, 8.0)

Table 3.5 Comparison results of the water pumping system.

k	$\hat{A}_2$	$\hat{A}_3$
1	(2.91, 3.9, 5.9)	(3.9, 5.9, 6.9)
2	(2.7, 3.71, 5.8)	(3.7, 5.7, 6.8)
⋮	⋮	⋮
75	(1.0, 2.0, 4.0)	(1.0, 3.0, 4.0)

Three types of variable x satisfy the triangle function (3.9) and the trapezoidal function (3.10)

$$x = \{(2, 3, 4), (3, 5, 6), (1, 3, 5, 7)\}. \tag{3.134}$$

The corresponding data related to R are

$$R = \{(13, 43, 90), (24, 99, 180), (6, 43, 99, 237)\}. \tag{3.135}$$

The parameters satisfy

$$A = (3, 4, 6), B = (1, 4, 5), C = (2, 3, 4). \tag{3.136}$$

The approximation results are shown in Table 3.6. ∎

Table 3.6 Neural network approximation for the coefficients of the heat source.

t	A	B	C
1	(5.8, 6.8, 8.8)	(3.9, 6.9, 7.92)	(3.9, 4.7, 5.9)
2	(5.7, 6.7, 8.6)	(3.8, 6.8, 7.8)	(3.8, 4.5, 5.8)
3	(5.6, 6.6, 8.5)	(3.6, 6.6, 7.7)	(3.6, 4.4, 5.6)
⋮	⋮	⋮	⋮
95	(3.0, 4.0, 6.0)	(1.0, 4.0, 5.0)	(2.0, 3.0, 4.0)

3.4 Summary

In order to model an uncertain nonlinear system, fuzzy equations and dual fuzzy equations are used, which are in the form of linear-in-parameter. The uncertainty is represented by fuzzy numbers. Initially, it was proved that these fuzzy models have solutions under certain conditions. These conditions are controllability of the fuzzy control algorithms. By some special fuzzy operations, the dual fuzzy equations are transformed into two types of neural networks. Modified gradient descent algorithms are designed to train the neural networks such that the solutions (fuzzy controllers) are estimated by the neural networks.

Since normal modeling methods cannot be applied for fuzzy number and fuzzy equation directly, the fuzzy equation is transformed into a neural network. Then the gradient descent method is modified for fuzzy number updating, and a back-propagation learning rule is proposed for fuzzy equations. The upper bounds of the fuzzy modeling errors are proved. The approximation theory of crisp models is successfully extended to fuzzy equation model. The new methods are validated with some benchmark examples.

4

Modeling and Control Using Fuzzy Differential Equations

4.1 Introduction

In this chapter, a new model based on a Bernstein neural network is used, which has the good properties of the Bernstein polynomial for fuzzy differential equations (FDEs). Two types of neural networks are used: static and dynamic models, to approximate the solutions of FDEs. These numerical methods use the generalized differentiability of FDEs. The solutions of FDEs are substituted into four ordinary differential equations (ODEs). Then the corresponding Bernstein neural networks are applied. Furthermore, a new method based on the fuzzy Sumudu transform (FST) is used to obtain the approximate solutions of FDEs. Significant theorems are suggested in order to explain the properties of the FST. The FST reduces the FDE to an algebraic equation. A very important property of the FST is that it can solve the equation without resorting to a new frequency domain. By utilizing the proposed technique, the fuzzy boundary value problem can be resolved directly without determining a general solution [44]. Some numerical examples are proposed to show the effectiveness of the approximation methods using the Bernstein neural network and the FST.

4.2 Fuzzy Modeling with Fuzzy Differential Equations

The Bernstein polynomial has good uniform approximation ability for continuous functions [62]. Also, it has innumerable drawing properties: homogeneous shape sustaining approximation, bona fide estimation, and low boundary bias. A very important property of the Bernstein polynomial is that it generates a smooth estimate for equal distance knots [60]. This property is suitable for FDE approximation [99].

In reference to fuzzy or interval arithmetic, the equation $x = y \oplus z$ is not equivalent to the phase $z = x \ominus y = x \oplus (-1)y$ or to $y = x \ominus z = x \oplus (-1)z$

Modeling and Control of Uncertain Nonlinear Systems with Fuzzy Equations and Z-Number,
First Edition. Wen Yu and Raheleh Jafari.

and this is a major factor for introducing the following Hukuhara difference (H-difference) [117].

Definition 4.1 (Hukuhara difference) *The definition of H-difference is suggested by* $x \ominus_H y = z \iff x = y \oplus z$; *if* $x \ominus_H y$ *exists, its α-level is* $[x \ominus_H y]^\alpha = [\underline{x}^\alpha - \underline{y}^\alpha, \overline{x}^\alpha - \overline{y}^\alpha]$. *Precisely,* $x \ominus_H y = 0$ *but* $x \ominus y \neq 0$.

Definition 4.2 (Generalized Hukuhara difference) [41] *The generalized Hukuhara difference (gH-difference) of two fuzzy variables x and y is illustrated as*

$$x \ominus_{gH} y = z \iff \begin{cases} 1) & x = y \oplus z \\ 2) & y = x \oplus (-1)z \end{cases} . \tag{4.1}$$

It is convenient to display that (1) and (2) below in combination are genuine if and only if z is a crisp number. With respect to the α-level the obtained results are $[x \ominus_{gH} y]^\alpha = [\min\{\underline{x}^\alpha - \underline{y}^\alpha, \overline{x}^\alpha - \overline{y}^\alpha\}, \max\{\underline{x}^\alpha - \underline{y}^\alpha, \overline{x}^\alpha - \overline{y}^\alpha\}]$ *and if* $x \ominus_{gH} y$ *and* $x \ominus_H y$ *hold,* $x \ominus_H y = x \ominus_{gH} y$. *The conditions for the existence of* $z = x \ominus_{gH} y \in E$ *are*

$$(1) \begin{cases} \underline{z}^\alpha = \underline{x}^\alpha - \underline{y}^\alpha \quad \text{and} \quad \overline{z}^\alpha = \overline{x}^\alpha - \overline{y}^\alpha \\ \text{with} \quad \underline{z}^\alpha \quad \text{increasing} \quad \overline{z}^\alpha \quad \text{decreasing} \quad \underline{z}^\alpha \leq \overline{z}^\alpha \end{cases}$$

$$(2) \begin{cases} \underline{z}^\alpha = \overline{x}^\alpha - \overline{y}^\alpha \quad \text{and} \quad \overline{z}^\alpha = \underline{x}^\alpha - \underline{y}^\alpha \\ \text{with} \quad \underline{z}^\alpha \quad \text{increasing} \quad \overline{z}^\alpha \quad \text{decreasing} \quad \underline{z}^\alpha \leq \overline{z}^\alpha \end{cases} \tag{4.2}$$

where $\forall \alpha \in [0, 1]$

Definition 4.3 (α−level of fuzzy function) *The α-level of fuzzy valued function* $F : [0, a] \to E$ *is*

$$F(x, \alpha) = [\underline{F}(x, \alpha), \overline{F}(x, \alpha)] \tag{4.3}$$

where $x \in E$, *for each* $\alpha \in [0, 1]$.

With the definition of gH-difference, the gH-derivative of F at x_0 is expressed as

$$\frac{d}{dt} F(x_0) = \lim_{h \to 0} \frac{1}{h} [F(x_0 + h) \ominus_{gH} F(x_0)]. \tag{4.4}$$

In (4.4), $F(x_0 + h)$ and $F(x_0)$ show a similar style with x and y, respectively, included in (4.1).

The following FDE is used

$$\frac{d}{dt} x = f(x, t) \tag{4.5}$$

where x is the fuzzy variable $x \in E$, $\frac{d}{dt}x$ is the fuzzy derivative (see the H-difference) of x, and $f(x,t)$ is a fuzzy function. It is clear that the fuzzy function $f(x, u)$ is the mapping $f : [0, \zeta] \times E \rightarrow E$, where $\zeta \in R$.

Let us consider the FDE (4.5) where $f : [0, \zeta] \times E \rightarrow E$. If the α-level (3.12) applies to $f(x, t)$ in (4.5), then two functions, $\underline{f}[t, \underline{x}(\zeta, \alpha), \overline{x}(\zeta, \alpha)]$ and $\overline{f}[t, \underline{x}(\zeta, \alpha), \overline{x}(\zeta, \alpha)]$, can be obtained.

The fuzzy differential Equation (4.5) can be equivalent to the following four ODEs

$$
(1) \begin{cases} \frac{d}{dt}\underline{x} = \underline{f}[t, \underline{x}(\zeta, \alpha), \overline{x}(\zeta, \alpha)] \\ \frac{d}{dt}\overline{x} = \overline{f}[t, \underline{x}(\zeta, \alpha), \overline{x}(\zeta, \alpha)] \end{cases}
$$

$$
(2) \begin{cases} \frac{d}{dt}\underline{x} = \overline{f}[t, \underline{x}(\zeta, \alpha), \overline{x}(\zeta, \alpha)] \\ \frac{d}{dt}\overline{x} = \underline{f}[t, \underline{x}(\zeta, \alpha), \overline{x}(\zeta, \alpha)] \end{cases} . \tag{4.6}
$$

if the following two conditions are satisfied [37]: $\underline{f}$ and $\overline{f}$ are "equivalent continuous", and $\underline{f}$ and $\overline{f}$ satisfy the Lipschitz conditions

$$
|\overline{f}[t, \underline{x}(\zeta, \alpha_1), \overline{x}(\zeta, \alpha_1)] - \overline{f}[t, \underline{x}(\zeta, \alpha_2), \overline{x}(\zeta, \alpha_2)]| \leq L_1 |\alpha_1 - \alpha_2|
$$
$$
|\underline{f}[t, \underline{x}(\zeta, \alpha_1), \overline{x}(\zeta, \alpha_1)] - \underline{f}[t, \underline{x}(\zeta, \alpha_2), \overline{x}(\zeta, \alpha_2)]| \leq L_2 |\alpha_1 - \alpha_2| \tag{4.7}
$$

where L_1 and L_2 are positive constants.

Definition 4.4 (Strongly generalized differentiable) *Suppose $f : [a_1, a_2] \rightarrow E$ and $t_0 = [a_1, a_2]$. f is a strongly generalized differentiable at t_0, if for all $h > 0$ it is adequately minute, $\frac{d}{dt}f(t_0) \in E$ exists in such a manner that*
(i) $\exists f(t_0 + h) \ominus_H f(t_0), f(t_0) \ominus_H f(t_0 - h)$, and

$$
\lim_{h \to 0^+} \frac{f(t_0 + h) \ominus_H f(t_0)}{h} = \lim_{h \to 0^+} \frac{f(t_0) \ominus_H f(t_0 - h)}{h} = \frac{d}{dt}f(t_0) \tag{4.8}
$$

or (ii) $\exists f(t_0) \ominus_H f(t_0 + h), f(t_0 - h) \ominus_H f(t_0)$, and

$$
\lim_{h \to 0^+} \frac{f(t_0) \ominus_H f(t_0 + h)}{(-h)} = \lim_{h \to 0^+} \frac{f(t_0 - h) \ominus_H f(t_0)}{(-h)} = \frac{d}{dt}f(t_0). \tag{4.9}
$$

Definition 4.5 (Integrable fuzzy function) *The function $f : [a_1, a_2] \rightarrow E$ is integrable on $[a_1, a_2]$ if it satisfies in the relation*

$$
\int_{a_1}^{\infty} f(t)dt = \left(\int_{a_1}^{\infty} \underline{f}(t, \alpha)dt, \int_{a_1}^{\infty} \overline{f}(t, \alpha)dt \right). \tag{4.10}
$$

If $f(t)$ is a fuzzy value function and $q(t)$ is a fuzzy Riemann integrable on $[a_1, \infty]$ so $f(t) \oplus q(t)$ can be a fuzzy Riemann integrable on $[a_1, \infty]$. Therefore,

$$
\int_{a_1}^{\infty} (f(t) \oplus q(t))dt = \int_{a}^{\infty} f(t)dt \oplus \int_{a}^{\infty} q(t)dt. \tag{4.11}
$$

For all $\alpha \in [0,1]$, the following results can be extracted [56].

i) If f is (i)-differentiable, so $\underline{f}(t,\alpha)$ and $\overline{f}(t,\alpha)$ are differentiable functions, moreover $\frac{d}{dt}f(t) = (\frac{d}{dt}\underline{f}(t,\alpha), \frac{d}{dt}\overline{f}(t,\alpha))$.

ii) If f is (ii)-differentiable, so $\underline{f}(t,\alpha)$ and $\overline{f}(t,\alpha)$ are differentiable functions, moreover $\frac{d}{dt}f(t) = (\frac{d}{dt}\overline{f}(t,\alpha), \frac{d}{dt}\underline{f}(t,\alpha))$.

Definition 4.6 (Fuzzy Sumudu transform) *Suppose $f(t)$ is a continuous fuzzy value function, and $f(Bt) \odot e^{-t}$ is an improper fuzzy Riemann integrable on $[0,\infty)$. Accordingly, $\int_0^\infty f(Bt) \odot e^{-t}dt$ is expressed as an FST and it is defined by $\Omega(B) = S[f(t)] = \int_0^\infty f(Bt) \odot e^{-t}dt$, where $0 \leq B < K, K \geq 0$, and e^{-t} is real valued function.*

4.3 Existence of a Solution

Theorem 4.1 *If the fuzzy function f and its derivative $\frac{\partial f}{\partial x}$ are on the rectangle $[-p,p] \times [-q,q]$, where $p,q \in E$, E is a fuzzy set there exists an unique fuzzy solution for the following FDE*

$$\frac{d}{dt}x = f(t,x), \quad x(t_0) = x_0 \tag{4.12}$$

for all $t \in (-b,b), b \leq p$.

Proof: Picard's iteration technique [43] is utilized to develop a sequence of fuzzy functions $\varphi_n(t)$ as

$$\varphi_{n+1}(t) = \varphi_0 \oplus \int_0^t f(s,\varphi_n(s))ds$$

$$= \varphi_0 \ominus_H (-1) \int_0^t f(s,\varphi_n(s))ds. \tag{4.13}$$

Initially, it should be validated that $\varphi_n(t)$ is continuous and exists for all n. Obviously, if $\varphi_n(t)$ exists then $\varphi_{n+1}(t)$ also exists as

$$\varphi_{n+1}(t) = \varphi_0 \oplus \int_0^t f(s,\varphi_n(s))ds$$

$$= \varphi_0 \ominus_H (-1) \int_0^t f(s,\varphi_n(s))ds. \tag{4.14}$$

Since f is continuous, so there exist $N \in E$ such that $|f(t,x)| \leq N$ for all $t \in [-p,p]$, and all $x \in [-q,q]$. If $t \in [-b,b]$ for $b \leq \min(q/N,p)$, then it is possible that

$$\| \varphi_{n+1} \ominus \varphi_0 \| = \left\| \int_0^t f(s,\varphi_n(s))ds \right\| \leq N|t| \leq Nb \leq q. \tag{4.15}$$

This validates that $\varphi_{n+1}(t)$ obtains values in $[-q, q]$, because

$$\varphi_n(t) = \sum_{k=1}^{n} (\varphi_n(t) \ominus \varphi_{n-1}(t)) \tag{4.16}$$

for any $\gamma < 1$, $t \in (-b, b)$ is selected in such a manner that $|\varphi_k(t) \ominus \varphi_{k-1}(t)| \leq \gamma^k$ for all k. This signifies that there exists $\gamma < 1$ [107]

$$|\varphi_k(t) \ominus \varphi_{k-1}(t)| \leq \gamma^k. \tag{4.17}$$

From the mean value theorem [160],

$$\varphi_k(t) \ominus \varphi_{k-1}(t) = \int_0^t [f(s, \varphi_{k-1}(s)) \ominus f(s, \varphi_{k-2}(s))] ds. \tag{4.18}$$

Applying the mean value theorem to the fuzzy function $h(x) = f(s, x)$ at the two points $\varphi_{k-1}(s)$ and $\varphi_{k-2}(s)$,

$$h(\varphi_{k-1}(s)) \ominus h(\varphi_{k-2}(s)) = h'(\psi_k(s))(\varphi_{k-1}(s)) \ominus \varphi_{k-2}(s)). \tag{4.19}$$

Taking into consideration $h'(x) = \frac{\partial f}{\partial x}$, the following is obtained

$$\varphi_k(t) \ominus \varphi_{k-1}(t) = \int_0^t \frac{\partial f}{\partial x}(s, \psi_k(s))(\varphi_{k-1}(s) \ominus \varphi_{k-2}(s)) ds. \tag{4.20}$$

Because $|\varphi_{k-1}(s) \ominus \varphi_{k-2}(s)| \leq \gamma^{k-1}$ for $s \leq t$ and $b < \gamma/N$, by substituting the above relation in (4.20) and bounding $\frac{\partial f}{\partial x}$ by N the following is obtained

$$|\varphi_k(t)) \ominus \varphi_{k-1}(t)| \leq \int_0^t N\gamma^{k-1} ds = Nt\gamma^{k-1} \leq Nb\gamma^{k-1}. \tag{4.21}$$

In order to validate that x is continuous, it is necessary to show that for any given $\epsilon > 0$ there exists $\delta > 0$ in such a manner that $|t_2 - t_1| < \delta$ implies $|\varphi(t_2) \ominus \varphi(t_1)| < \epsilon$. For notation convenience, it has been supposed that $t_1 < t_2$. It follows that

$$\varphi(t_2) \ominus \varphi(t_1) = \lim_{n \to \infty} \varphi_n(t_2) \ominus \lim_{n \to \infty} \varphi_n(t_1)$$

$$= \lim_{n \to \infty} (\varphi_n(t_2) \ominus \varphi_n(t_1)) = \lim_{n \to \infty} \int_{t_1}^{t_2} f(s, \varphi_n(s)) ds. \tag{4.22}$$

There exists N in such a manner that $|f(s, x)| \leq N$. Hence

$$|\varphi(t_2) \ominus \varphi(t_1)| \leq \int_{t_1}^{t_2} N ds = N |t_2 - t_1| \leq N\delta. \tag{4.23}$$

Thus, by selecting $\delta < \epsilon/N$ it is observed that $|\varphi(t_2) \ominus \varphi(t_1)| < \epsilon$. So $\lim_{n \to \infty} \varphi_n(t)$ exists for all t.

Now it should be demonstrated that $\lim_{n\to\infty}\varphi_n(t)$ is continuous. Since

$$\varphi(t) = \lim_{n\to\infty}\varphi_n(t) = \lim_{n\to\infty}\int_0^t f(s,\varphi_{n-1}(s))ds$$

$$= \int_0^t \lim_{n\to\infty} f(s,\varphi_{n-1}(s))ds = \int_0^t f(s,\lim_{n\to\infty}\varphi_{n-1}(s))ds \tag{4.24}$$

where the last step (moving the limit inside the function) follows from the fact that f is continuous in each variable. Hence it is clear that

$$\varphi(t) = \int_0^t f(s,\varphi(s))ds \tag{4.25}$$

because all functions are continuous,

$$\frac{\mathrm{d}}{\mathrm{d}t}\varphi = f(s,\varphi(t)). \tag{4.26}$$

If there exists another solution $\phi(t)$,

$$\varphi(t) \ominus \phi(t) = \int_0^t (f(s,\varphi(t)) \ominus f(s,\phi(t)))ds. \tag{4.27}$$

Since the two functions are different, there exists $\epsilon > 0$ and $|\varphi(t)\ominus\phi(t)| > \epsilon$. The following is defined

$$m = \max_{0\leq t\leq b} |\varphi(t)\ominus\phi(t)| \tag{4.28}$$

N is the bound for $\frac{\partial f}{\partial x}$. Utilizing the mean value theorem,

$$|\varphi(t)\ominus\phi(t)| \leq \int_0^t N\,|\varphi(t)-\phi(t)|\,ds \leq N\,|t|\,m \leq Nbm \tag{4.29}$$

If $b < \epsilon/2mN$, it signifies that for all $t < b$, $|\varphi(t)-\phi(t)| < \epsilon/2$, indicating that the least difference is ϵ. So there exists a unique fuzzy solution. ∎

Theorem 4.2 *Assume the following FDE*

$$\frac{\mathrm{d}}{\mathrm{d}t}x = f(t,x) \tag{4.30}$$

where $f \in \bar{J}_{ab}, \bar{J}_{ab}$ is the set of linear strongly bounded operators, for every operator f there exists a function $\tau \in L([a,b]; E_+)$ such that $|f(v)(t)| \leq \tau(t)\,\|v\|_G, t \in [a,b]$ and $v \in G([a,b]; E)$, and there exists $f_0, f_1 \in \varphi_{ab}$, where φ_{ab} is a set of linear operators $f \in \bar{J}_{ab}$ from the set $G([a,b]; E_+)$ to the set $L([a,b]; E_+)$, such that

$$|\underline{f}(t,\underline{v},\bar{v}) + \underline{f}_1(t,\underline{v},\bar{v})| \leq \underline{f}_0(t,|\underline{v}|,|\bar{v}|), \quad t \in [a,b]$$

$$|\bar{f}(t,\underline{v},\bar{v}) + \bar{f}_1(t,\underline{v},\bar{v})| \leq \bar{f}_0(t,|\underline{v}|,|\bar{v}|), \quad t \in [a,b] \tag{4.31}$$

then (4.12) has a unique solution.

Proof: If x is a solution of (4.30) and $-\frac{1}{2}f_1 \in J_{ab}(a)$,

$$\frac{d}{dt}\beta = -\frac{1}{2}f_1(t,\beta) \oplus f_0(t,|x|) \oplus \frac{1}{2}f_1(t,|x|) \tag{4.32}$$

contains a unique solution β. Moreover, as $f_0, f_1 \in \varphi_{ab}$

$$\underline{\beta}(t) \geq 0, \quad t \in [a,b]$$
$$\overline{\beta}(t) \geq 0, \quad t \in [a,b]. \tag{4.33}$$

According to (4.31) and the condition $f_1 \in \varphi_{ab}$, from (4.32) the following is obtained

$$\frac{d}{dt}\underline{\beta} \geq -\frac{1}{2}\underline{f}_1(t,\underline{\beta},\overline{\beta}) + \underline{f}(t,\underline{x},\overline{x}) + \frac{1}{2}\underline{f}_1(t,\underline{x},\overline{x})$$
$$\frac{d}{dt}\overline{\beta} \geq -\frac{1}{2}\overline{f}_1(t,\underline{\beta},\overline{\beta}) + \overline{f}(t,\underline{x},\overline{x}) + \frac{1}{2}\overline{f}_1(t,\underline{x},\overline{x}) \tag{4.34}$$

thus $t \in [a,b]$

$$\frac{d}{dt}(-\underline{\beta}) \leq -\frac{1}{2}\underline{f}_1(t,-\underline{\beta},-\overline{\beta}) + \underline{k}(t,\underline{x},\overline{x}) + \frac{1}{2}\underline{k}_1(t,\underline{x},\overline{x})$$
$$\frac{d}{dt}(-\overline{\beta}) \leq -\frac{1}{2}\overline{f}_1(t,-\underline{\beta},-\overline{\beta}) + \overline{f}(t,\underline{x},\overline{x}) + \frac{1}{2}\overline{f}_1(t,\underline{x},\overline{x}). \tag{4.35}$$

The last two inequalities are due to the assumption $-\frac{1}{2}f_1 \in J_{ab}(a)$

$$|\underline{x}(t)| \leq \underline{\beta}(t) \quad t \in [a,b]$$
$$|\overline{x}(t)| \leq \overline{\beta}(t) \quad t \in [a,b]. \tag{4.36}$$

According to (4.36) and the conditions $f_0, f_1 \in \varphi_{ab}$, (4.32) results in

$$\frac{d}{dt}\underline{\beta} \leq \underline{f}_0(t,\underline{\beta},\overline{\beta}), \quad t \in [a,b]$$
$$\frac{d}{dt}\overline{\beta} \leq f_0(t,\underline{\beta},\overline{\beta}), \quad t \in [a,b]. \tag{4.37}$$

As $f_0 \in J_{ab}(a)$, the last inequality with $\beta(a) = 0$ yields $\underline{\beta}(t) \leq 0$ and $\overline{\beta}(t) \leq 0$ for $t \in [a,b]$. Equation (4.33) implies $\beta \equiv 0$. Thus based on (4.36), $x \equiv 0$ holds. ∎

Theorem 4.3 *Assume $f : R \times E \to E$ is taken to be a continuous fuzzy function. If $t_0 \in R$, the FDE (4.5) is incorporated with two solutions: (i)-differentiable, and (ii)-differentiable. Hence the successive iterations*

$$x_{n+1}(t) = x_0 + \int_{t_0}^{t} f(t,x_n(t))dt, \quad \forall t \in [t_0,t_1] \tag{4.38}$$

and

$$x_{n+1}(t) = x_0 \ominus_H (-1)\int_{t_0}^{t} f(t,x_n(t))dt, \quad \forall t \in [t_0,t_1] \tag{4.39}$$

approaches the two solutions sequentially.

The proof of this theorem is straightforward, see [40].

Theorem 4.4 *Suppose $\frac{d}{dt}f(t)$ is a fuzzy value integrable function, as well as $f(t)$ being the primitive of $\frac{d}{dt}f(t)$ on $[0, \infty)$. Therefore,*

$$S\left[\frac{d}{dt}f(t)\right] = \frac{1}{B} \odot S[f(t)] \ominus \left(\frac{1}{B} \odot [f(0)]\right) \tag{4.40}$$

where f is considered to be (i)-differentiable, or

$$S\left[\frac{d}{dt}f(t)\right] = \frac{-1}{B} \odot [f(0)] \ominus \left(\frac{-1}{B} \odot S[f(t)]\right) \tag{4.41}$$

where f is considered to be (ii)-differentiable.

Proof: For arbitrary fixed $\alpha \in [0, 1]$ the following is obtained

$$\frac{1}{B} \odot S[f(t)] \ominus \left(\frac{1}{B} \odot f(0)\right)$$
$$= \left(\frac{1}{B}S[\underline{f}(t, \alpha)] - \frac{1}{B}S[\underline{f}(0, \alpha)], \frac{1}{B}S[\overline{f}(t, \alpha)] - \frac{1}{B}S[\overline{f}(0, \alpha)]\right). \tag{4.42}$$

Furthermore,

$$S\left[\frac{d}{dt}\overline{f}(t, \alpha)\right] = \frac{1}{B}S[\overline{f}(t, \alpha)] - \frac{1}{B}[\overline{f}(0, \alpha)]$$
$$S\left[\frac{d}{dt}\underline{f}(t, \alpha)\right] = \frac{1}{B}S[\underline{f}(t, \alpha)] - \frac{1}{B}[\underline{f}(0, \alpha)]. \tag{4.43}$$

Hence,

$$\frac{1}{B} \odot S[f(t)] \ominus \left(\frac{1}{B} \odot f(0)\right) = \left(S\left[\frac{d}{dt}\underline{f}(t, \alpha)\right], S\left[\frac{d}{dt}\overline{f}(t, \alpha)\right]\right). \tag{4.44}$$

If f is considered to be (i)-differentiable, then

$$\frac{1}{B} \odot S[f(t)] \ominus \left(\frac{1}{B} \odot f(0)\right) = S\left[\frac{d}{dt}f(t)\right]. \tag{4.45}$$

Let f is (ii)-differentiable. For arbitrary fixed $\alpha \in [0, 1]$ the following is obtained

$$\frac{-1}{B} \odot [f(0)] \ominus \left(\frac{-1}{B} \odot S[f(t)]\right)$$
$$= \left(\frac{-1}{B}\overline{f}(0, \alpha) + \frac{1}{B}S[\overline{f}(t, \alpha)], \frac{-1}{B}\underline{f}(0, \alpha) + \frac{1}{B}S[\underline{f}(t, \alpha)]\right). \tag{4.46}$$

The above equation can be written as the following relation

$$\frac{-1}{B} \odot [f(0)] \ominus \left(\frac{-1}{B} \odot S[f(t)]\right)$$
$$= \left(\frac{1}{B}S[\overline{f}(t, \alpha)] - \frac{1}{B}\overline{f}(0, \alpha), \frac{1}{B}S[\underline{f}(t, \alpha)] - \frac{1}{B}\underline{f}(0, \alpha)\right). \tag{4.47}$$

The following is obtained

$$\mathbf{S}\left[\frac{\mathrm{d}}{\mathrm{d}t}\overline{f}(t,\alpha)\right] = \frac{1}{B}\mathbf{S}[\overline{f}(t,\alpha)] - \frac{1}{B}\overline{f}(0,\alpha)$$

$$\mathbf{S}\left[\frac{\mathrm{d}}{\mathrm{d}t_-}\underline{f}(t,\alpha)\right] = \frac{1}{B}\mathbf{S}[\underline{f}(t,\alpha)] - \frac{1}{B}\underline{f}(0,\alpha). \tag{4.48}$$

So

$$\left(\frac{-1}{B}f(0)\right) \ominus \left(\frac{-1}{B} \odot \mathbf{S}[f(t)]\right) = \left(\mathbf{S}[\overline{f}(t,\alpha)], \mathbf{S}\left[\frac{\mathrm{d}}{\mathrm{d}t_-}\underline{f}(t,\alpha)\right]\right). \tag{4.49}$$

Hence

$$\left(\frac{-1}{B}f(0)\right) \ominus \left(\frac{-1}{B} \odot \mathbf{S}[f(t)]\right) = \mathbf{S}\left(\left[\frac{\mathrm{d}}{\mathrm{d}t}\overline{f}(t,\alpha)\right], \left[\frac{\mathrm{d}}{\mathrm{d}t_-}\underline{f}(t,\alpha)\right]\right). \tag{4.50}$$

Since f is (ii)-differentiable, therefore,

$$\left(\frac{-1}{B}f(0)\right) \ominus \left(\frac{-1}{B} \odot \mathbf{S}[f(t)]\right) = \mathbf{S}\left[\frac{\mathrm{d}}{\mathrm{d}t'}f(t)\right]. \tag{4.51}$$

∎

Theorem 4.5 *Taking into consideration that Sumudu transform is a linear transformation, so if $f_1(t)$ and $f_2(t)$ be continuous fuzzy valued functions, moreover k_1 as well as k_2 be constant, therefore the following relation can be obtained*

$$\mathbf{S}[(k_1 \odot f_1(t)) \oplus (k_2 \odot f_2(t))] = (k_1 \odot \mathbf{S}[f_1(t)]) \oplus (k_2 \odot \mathbf{S}[f_2(t)]). \tag{4.52}$$

Proof: Since

$$\mathbf{S}[(k_1 \odot f_1(t)) \oplus (k_2 \odot f_2(t))]$$

$$= \int_0^\infty \left(k_1 \odot f_1(Bt) \oplus k_2 \odot f_2(Bt)\right) \odot e^{-t}\mathrm{d}t$$

$$= \int_0^\infty k_1 \odot f_1(Bt) \odot e^{-t}\mathrm{d}t \oplus \int_0^\infty k_2 \odot f_2(Bt) \odot e^{-t}\mathrm{d}t$$

$$= k_1 \odot \left(\int_0^\infty f_1(Bt) \odot e^{-t}\mathrm{d}t\right) \oplus k_2 \odot \left(\int_0^\infty f_2(Bt) \odot e^{-t}\mathrm{d}t\right)$$

$$= k_1 \odot \mathbf{S}[f_1(t)] \oplus k_2 \odot \mathbf{S}[f_2(t)]. \tag{4.53}$$

Therefore, the following is concluded

$$\mathbf{S}[(k_1 \odot f_1(t)) \oplus (k_2 \odot f_2(t))] = (k_1 \odot \mathbf{S}[f_1(t)]) \oplus (k_2 \odot \mathbf{S}[f_2(t)]). \tag{4.54}$$

∎

Based on the Theorem 4.5 the following relation is obtained

$$\int_0^\infty f(Bt) \odot e^{-t}\mathrm{d}t = \left(\int_0^\infty \underline{f}(Bt,\alpha)e^{-t}\mathrm{d}t, \int_0^\infty \overline{f}(Bt,\alpha)e^{-t}\mathrm{d}t\right). \tag{4.55}$$

Let

$$\mathbf{S}[\underline{f}(t,\alpha)] = \int_0^\infty \underline{f}(Bt,\alpha)e^{-t}dt$$

$$\mathbf{S}[\overline{f}(t,\alpha)] = \int_0^\infty \overline{f}(Bt,\alpha)e^{-t}dt \tag{4.56}$$

hence,

$$\mathbf{S}[f(t)] = (\mathbf{S}[\underline{f}(t,\alpha), \mathbf{S}\overline{f}(t,\alpha)]). \tag{4.57}$$

Lemma 4.1 *Assume that $f(t)$ is a continuous fuzzy value function on $[0,\infty)$, also $\gamma \geq 0$, thus*

$$\mathbf{S}[\gamma \odot f(t)] = \gamma \odot \mathbf{S}[f(t)]. \tag{4.58}$$

Proof: Fuzzy Sumudu transform $\gamma \odot f(t)$ is defined as

$$\mathbf{S}[\gamma \odot f(t)] = \int_0^\infty \gamma \odot f(Bt) \odot e^{-t}dt. \tag{4.59}$$

Furthermore,

$$\int_0^\infty \gamma \odot f(Bt) \odot e^{-t}dt = \gamma \odot \int_0^\infty f(Bt) \odot e^{-t}dt. \tag{4.60}$$

Therefore,

$$\mathbf{S}[\gamma \odot f(t)] = \gamma \odot \mathbf{S}[f(t)]. \tag{4.61}$$

∎

Lemma 4.2 *Assume that $f_1(t)$ is a continuous fuzzy value function, and $f_2(t) \geq 0$. Furthermore, if $(f_1(t) \odot f_2(t)) \odot e^{-t}$ be improper fuzzy Riemann integrable on $[0,\infty)$, then*

$$\int_0^\infty (f_1(Bt) \odot f_2(Bt)) \odot e^{-t}dt$$

$$= \left(\int_0^\infty f_2(Bt)\underline{f}_1(Bt,\alpha)e^{-t}dt, \int_0^\infty f_2(Bt)\overline{f}_1(Bt,\alpha)e^{-t}dt \right). \tag{4.62}$$

Theorem 4.6 *Suppose $f(t)$ is a continuous fuzzy value function, also $\mathbf{S}[f(t)] = D(B)$, therefore,*

$$\mathbf{S}[e^{a_1t} \odot f(t)] = \frac{1}{1-a_1B}D\left(\frac{B}{1-a_1B}\right) \tag{4.63}$$

where e^{a_1t} is considered to be a real value function, also $1 - a_1B > 0$.

Proof: The following relation is obtained

$$S[e^{a_1 t} \odot f(t)] = \int_0^\infty e^{a_1 Bt} e^{-t} f(Bt) dt$$

$$= \left(\int_0^\infty e^{-(1-a_1 B)t} \underline{f}(Bt, \alpha) dt, \int_0^\infty e^{-(1-a_1 B)t} \overline{f}(Bt, \alpha) dt \right).$$

(4.64)

Suppose $z = 1 - a_1 Bt$, then

$$S[e^{a_1 t} \odot f(t)]$$

$$= \frac{1}{1-a_1 B} \left(\int_0^\infty \underline{f}\left(\frac{Bz}{1-a_1 B}, \alpha\right) e^{-z} dz, \int_0^\infty \overline{f}\left(\frac{Bz}{1-a_1 B}, \alpha\right) e^{-z} dz \right)$$

$$= \left\{ \frac{1}{1-a_1 B} \underline{D}\left(\frac{B}{1-a_1 B}\right), \frac{1}{1-a_1 B} \overline{D}\left(\frac{B}{1-a_1 B}\right) \right\} = \frac{1}{1-a_1 B} D\left(\frac{B}{1-a_1 B}\right).$$

(4.65)

∎

4.4 Solution Approximation using Bernstein Neural Networks

In general, it is difficult to solve four ODE in (4.6). Here, a special neural network, Bernstein neural network, is used to approximate the solutions of the FDE (4.5).

The two variables Bernstein series polynomial can be written as follow

$$B(x_1, x_2) = \sum_{i=0}^N \sum_{j=0}^M \binom{N}{i} \binom{M}{j}$$

$$W_{i,j} x_{1i} (T - x_{1i})^{N-i} x_{2j} (1 - x_{2j})^{M-j}$$

(4.66)

where $\binom{N}{i} = \frac{N!}{i!(N-i)!}$, $\binom{M}{j} = \frac{M!}{j!(M-j)!}$, $W_{i,j}$ is the coefficient. This polynomial can be considered as neural network. It has two inputs: x_{1i} and x_{2j}, and one output y,

$$y = \sum_{i=0}^N \sum_{j=0}^M \lambda_i \gamma_j W_{i,j} x_{1i} (T - x_{1i})^{N-i} x_{2j} (1 - x_{2j})^{M-j}$$

(4.67)

where $\lambda_i = \binom{N}{i}$, $\gamma_j = \binom{M}{j}$.

Now the Bernstein neural network (4.67) is used to approximate the solution of the FDE (4.5). Since the solution of (4.5) can be written as four ODE as (4.6), so the Bernstein neural network is designed in the form of (4.6).

It has been assumed that x_1 is time interval t, x_2 is the α-level as in (3.12). The solution of (4.5) in the form of the Bernstein neural network is

$$x_m(t, \alpha) = x_m(0, \alpha)$$

$$\oplus t \sum_{i=0}^{N} \sum_{j=0}^{M} \lambda_i \gamma_j W_{i,j} t_i (T - t_i)^{N-i} \alpha_j (1 - \alpha_j)^{M-j} \tag{4.68}$$

where $x_m(0, \alpha)$ is the initial condition of the solution.

By calculating the derivative of (4.67), the following is obtained

$$(1) \begin{cases} \frac{d}{dt}\underline{x}_m = C_1 + C_2 \\ \frac{d}{dt}\overline{x}_m = D_1 + D_2 \end{cases}$$

$$(2) \begin{cases} \frac{d}{dt}\underline{x}_m = C_1 + C_2 \\ \frac{d}{dt}\overline{x}_m = D_1 + D_2 \end{cases} \tag{4.69}$$

where $t \in [0, T]$, $\alpha \in [0, 1]$, $t_k = kh$, $h = \frac{T}{k}$, $k = 1, ..., N$, $\alpha_j = jh_1$, $h_1 = \frac{1}{M}$, $j = 1, ..., M$,

$$C_1 = \sum_{i=0}^{N} \sum_{j=0}^{M} \lambda_i \gamma_j \underline{W}_{i,j} t_i (T - t_i)^{N-i} \alpha_j (1 - \alpha_j)^{M-j}$$

$$D_1 = \sum_{i=0}^{N} \sum_{j=0}^{M} \lambda_i \gamma_j \overline{W}_{i,j} t_i (T - t_i)^{N-i} \alpha_j (1 - \alpha_j)^{M-j}$$

$$C_2 = t_k \sum_{i=0}^{N} \sum_{j=0}^{M} \lambda_i \gamma_j \underline{W}_{i,j} [it_{i-1,j} (T - t_i)^{N-i}$$
$$- (N - i) t_{i,j} (T - t_i)^{N-i-1}] \alpha_j^i (1 - \alpha_j)^{M-j}$$

$$D_2 = t_k \sum_{i=0}^{N} \sum_{j=0}^{M} \lambda_i \gamma_j \overline{W}_{i,j} [it_{i-1,j} (T - t_i)^{N-i}$$
$$- (N - i) t_{i,j} (T - t_i)^{N-i-1}] \alpha_j^i (1 - \alpha_j)^{M-j} \tag{4.70}$$

where $d = 0.02$, $E = 70 \times 10^9$. The exact solution is $x = 4$.

- Input unit:
$$o_1^1 = t, \quad o_2^1 = \alpha. \tag{4.71}$$

- The first hidden units:
$$o_{1,i}^2 = f_i^1(o_1^1), \quad o_{2,i}^2 = f_i^2(o_1^1)$$
$$o_{3,j}^2 = g_j^1(o_2^1), \quad o_{4,j}^2 = g_j^2(o_2^1). \tag{4.72}$$

- The second hidden units:
$$o_{1,i}^3 = o_{1,i}^2(o_{2,i}^2), \quad o_{2,j}^3 = o_{3,j}^2(o_{4,j}^2). \tag{4.73}$$

- The third hidden units:

$$o_{1,i}^4 = \lambda_i o_{1,i}^3, \quad o_{2,i'}^4 = \gamma_j o_{2,j}^3. \tag{4.74}$$

- The forth hidden units:

$$o_{i,j}^5 = o_{1,i}^4 o_{2,j}^4. \tag{4.75}$$

- Output unit:

$$N(t, \alpha) = \sum_{i=0}^{N} \sum_{j=0}^{M} (a_{i,j} o_{i,j}^5). \tag{4.76}$$

Here $f_i^1 = t^i$, $f_i^2 = (T - t)^{N-i}$, $\lambda_i = \frac{1}{T^N}\binom{N}{i}$, $g_{i'}^1 = \alpha^j$, $g_j^2 = (1 - \alpha)^{M-j}$, $\gamma_j = \binom{M}{j}$.

The training errors between (4.69) and (4.6) are defined as

$$(1) \begin{cases} \underline{e}_1 = C_1 + C_2 - \underline{f} \\ \overline{e}_1 = D_1 + D_2 - \overline{f} \end{cases}$$
$$(2) \begin{cases} \underline{e}_2 = C_1 + C_2 - \overline{f} \\ \overline{e}_2 = D_1 + D_2 - \underline{f} \end{cases}. \tag{4.77}$$

The standard back-propagation learning algorithm is utilized to update the weights with the above training errors

$$\underline{W}_{i,j}(k + 1) = \underline{W}_{i,j}(k) - \eta_1 \left(\frac{\partial \underline{e}_1^2}{\partial \underline{W}_{i,j}} + \frac{\partial \overline{e}_1^2}{\partial \underline{W}_{i,j}} \right)$$
$$\overline{W}_{i,j}(k + 1) = \overline{W}_{i,j}(k) - \eta_2 \left(\frac{\partial \underline{e}_2^2}{\partial \overline{W}_{i,j}} + \frac{\partial \overline{e}_2^2}{\partial \overline{W}_{i,j}} \right) \tag{4.78}$$

where η_1 and η_1 are positive learning rates. The momentum terms, $\gamma \Delta \underline{W}_{i,j}(k - 1)$ and $\gamma \Delta \overline{W}_{i,j}(k - 1)$, can be added to stabilized the training process.

The Bernstein neural network (4.68) shown in Figure 4.1 is feed-forward (static) neural network. A recurrent (dynamic) neural network can also be used to approximate the solution of (4.5).

The dynamic Bernstein neural network is

$$\begin{cases} \frac{d}{dt}\underline{x}_m(t, \alpha) = \underline{P}(t, \alpha)A(\underline{x}_m(t, \alpha), \overline{x}_m(t, \alpha)) + \underline{Q}(t, \alpha) \\ \frac{d}{dt}\overline{x}_m(t, \alpha) = \overline{P}(t, \alpha)A(\underline{x}_m(t, \alpha), \overline{x}_m(t, \alpha)) + \overline{Q}(t, \alpha) \end{cases} \tag{4.79}$$

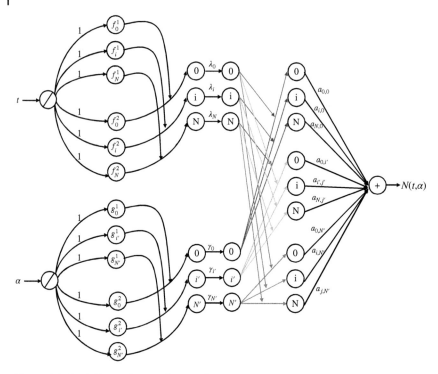

Figure 4.1 Static Bernstein neural network.

where $f(x, t)$ in (4.5) has the form of

$$f(t, x) = P(t)x + Q(t). \tag{4.80}$$

The structure of the dynamic Bernstein neural network is shown in Figure 4.2. The training errors in (4.77) are changed into

$$\text{(1)} \begin{cases} \underline{e}_1 = C_1 + C_2 - \underline{P}A(\underline{x}_m, \overline{x}_m) - \underline{Q} \\ \overline{e}_1 = D_1 + D_2 - \overline{P}A(\underline{x}_m, \overline{x}_m) - \overline{Q} \end{cases}$$

$$\text{(2)} \begin{cases} \underline{e}_2 = C_1 + C_2 - \overline{P}A(\underline{x}_m, \overline{x}_m) - \overline{Q} \\ \overline{e}_2 = D_1 + D_2 - \underline{P}A(\underline{x}_m, \overline{x}_m) - \underline{Q} \end{cases}. \tag{4.81}$$

The training algorithm can be the same as (4.78).

The learning process of the dynamic Bernstein neural network (4.79) is faster than the static Bernstein neural network (4.68). However, the robustness of (4.68) is better than (4.79), because the weights of the dynamic Bernstein neural network are difficult to converge.

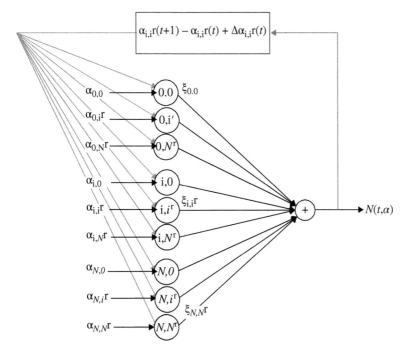

Figure 4.2 Dynamic Bernstein nerual network.

4.5 Solutions Approximation using the Fuzzy Sumudu Transform

Fuzzy initial and boundary value problems can be solved by utilizing the fuzzy Laplace transform [26]. In this section, the FST methodology is illustrated. By applying the FST methodology, the FDE is reduced to an algebraic equation. The main advantage of the FST is that it can solve the equation without resorting to a new frequency domain. The methodology of converting FDEs to an algebraic equation is expressed in [26].

Consider the following fuzzy initial value problem

$$
\begin{cases}
\frac{d}{dt}x(t) = f(t, x(t)), \\
x(0) = (\underline{x}(0, \alpha), \overline{x}(0, \alpha)), \quad 0 < \alpha \le 1.
\end{cases}
\tag{4.82}
$$

where $f(t, x(t))$ is a fuzzy function. The fuzzy function $f(t, x(t))$ is the mapping of $f : R \times E \to E$. By utilizing the FST method, the following is obtained

$$
S\left[\frac{d}{dt}x(t)\right] = S[f(t, x(t))].
\tag{4.83}
$$

The resolving process of (4.83) is base on the following cases.

Case 1: Assume that $\frac{d}{dt}x(t)$ is (i)-differentiable. Base on the Theorem 4.5 the following is extracted

$$\frac{d}{dt}x(t) = \left(\frac{d}{dt}\underline{x}(t,\alpha), \frac{d}{dt}\overline{x}(t,\alpha) \right) \tag{4.84}$$

$$S\left[\frac{d}{dt}x(t) \right] = \left(\frac{1}{B} \odot S[x(t)] \right) \ominus \frac{1}{B}x(0). \tag{4.85}$$

Equation (4.85) can be displayed as

$$\begin{cases} S[\underline{f}(t,x(t),\alpha)] = \frac{1}{B}S[\underline{x}(t,\alpha)] - \frac{1}{B}\underline{x}(0,\alpha) \\ S[\overline{f}(t,x(t),\alpha)] = \frac{1}{B}S[\overline{x}(t,\alpha)] - \frac{1}{B}\overline{x}(0,\alpha) \end{cases} \tag{4.86}$$

where

$$\begin{cases} \underline{f}(t,x(t),\alpha) = \min\{f(t,B)|B \in (\underline{x}(t,\alpha),\overline{x}(t,\alpha))\} \\ \overline{f}(t,x(t),\alpha) = \max\{f(t,B)|B \in (\underline{x}(t,\alpha),\overline{x}(t,\alpha))\}. \end{cases} \tag{4.87}$$

Accordingly, (4.87) can be solved on the basis of the following assumptions

$$S[\underline{x}(t,\alpha)] = U_1(B,\alpha) \tag{4.88}$$

$$S[\overline{x}(t,\alpha)] = U_2(B,\alpha) \tag{4.89}$$

where $U_1(B,\alpha)$, and $U_2(B,r)$ are the solutions of the (4.87). By applying inverse Sumudu transform, $\underline{x}(t,\alpha)$ and $\overline{x}(t,\alpha)$ are computed as

$$\underline{x}(t,\alpha) = S^{-1}[U_1(B,\alpha)] \tag{4.90}$$

$$\overline{x}(t,\alpha) = S^{-1}[U_2(B,\alpha)]. \tag{4.91}$$

Case 2: Assume that the $\frac{d}{dt}x(t)$ is (ii)-differentiable. Based on the Theorem 4.5 the following is extracted

$$\frac{d}{dt}x(t) = \left(\frac{d}{dt}\overline{x}(t,\alpha), \frac{d}{dt}\underline{x}(t,\alpha) \right) \tag{4.92}$$

$$S\left[\frac{d}{dt}x(t) \right] = \left(\frac{-1}{B} \odot x(0) \right) \ominus \left(\frac{-1}{B} \odot S[x(t)] \right). \tag{4.93}$$

Equation (4.93) can be displayed as

$$\begin{cases} S[\underline{f}(t,x(t),\alpha)] = \frac{1}{B}S[\underline{x}(t,\alpha)] - \frac{1}{B}\underline{x}(0,\alpha) \\ S[\overline{f}(t,x(t),\alpha)] = \frac{1}{B}S[\overline{x}(t,\alpha)] - \frac{1}{B}\overline{x}(0,\alpha) \end{cases} \tag{4.94}$$

where

$$\begin{cases} \underline{f}(t,x(t),\alpha) = \min\{f(t,B)|B \in (\underline{x}(t,\alpha),\overline{x}(t,\alpha))\} \\ \overline{f}(t,x(t),\alpha) = \max\{f(t,B)|B \in (\underline{x}(t,\alpha),\overline{x}(t,\alpha))\} \end{cases} \tag{4.95}$$

Accordingly, (4.95) can be resolved on the basis of the following assumptions

$$S(\underline{x}(t, \alpha) = V_1(B, \alpha)$$
$$S(\overline{x}(t, \alpha) = V_2(B, \alpha) \qquad (4.96)$$

where $V_1(B, \alpha)$, and $V_2(B, \alpha)$ are the solutions of the (4.95). By applying inverse Sumudu transform, $\underline{x}(t, \alpha)$ as well as $\overline{x}(t, \alpha)$ are computed as

$$\underline{x}(t, \alpha) = S^{-1}[V_1(B, \alpha)]$$
$$\overline{x}(t, \alpha) = S^{-1}[V_2(B, \alpha)]. \qquad (4.97)$$

4.6 Simulations

In this section, several real applications are used to show how to use the Bernstein neural networks to approximate the solutions of the FDEs.

Example 4.1 The vibration mass shown in Figure 2.1 has a very simple model

$$\frac{d}{dt}x(t) = \frac{k}{m}x(t) \qquad (4.98)$$

where the spring constant $k = 1$, the mass m are changeable in $(0.75, 1.125)$, so the position state $x(t)$ has some uncertainties, the ODE (4.98) can be formed into a FDE. It has the same form as (4.98), only $x(t)$ is considered to be a fuzzy variable. If the initial position is $x(0) = (0.75 + 0.25\alpha, 1.125 - 0.125\alpha)$, $\alpha \in [0, 1]$, then the exact solutions of the FDE (4.98) are [78]

$$x(t, \alpha) = [(0.75 + 0.25\alpha)e^t, (1.125 - 0.125\alpha)e^t] \qquad (4.99)$$

where $t \in [0, 1]$. Now the static Bernstein neural network (4.68), noted as SNN, is used to approximate the solution (4.99)

$$\begin{cases} \underline{x}_m(t, \alpha) = (0.75 + 0.25\alpha) \\ \qquad + t\sum\limits_{i=0}^{N}\sum\limits_{j=0}^{M} \lambda_i \gamma_j \underline{W}_{i,j} t_i (T - t_i)^{N-i} \alpha_j (1 - \alpha_j)^{M-j} \\ \overline{x}_m(t, \alpha) = (1.125 - 0.125\alpha) \\ \qquad + t\sum\limits_{i=0}^{N}\sum\limits_{j=0}^{M} \lambda_i \gamma_j \overline{W}_{i,j} t_i (T - t_i)^{N-i} \alpha_j (1 - \alpha_j)^{M-j} \end{cases} \qquad (4.100)$$

A dynamic Bernstein neural network (4.79), noted as DNN, is also used to approximate the solutions. The learning rates are $\eta = 0.01$, $\gamma = 0.01$ To compare the results, the other two popular methods: Max-Min Euler method and Average Euler method [185] are used. The comparison results are shown in Table 4.1 and Table 4.2. Corresponding solution plots are shown in Figure 4.3. ∎

Table 4.1 Solutions of different methods.

α	Exact solution	SNN	DNN	Max-Min Euler	Average Euler
0	[2.038,3.058]	[1.971,3.004]	[1.99,3.03]	[1.945,2.918]	[2.244,2.619]
0.1	[2.106,3.024]	[2.030,2.941]	[2.059,2.972]	[2.01,2.885]	[2.279,2.616]
0.2	[2.174,2.990]	[2.105,2.913]	[2.128,2.939]	[2.075,2.853]	[2.314,2.614]
0.3	[2.242,2.956]	[2.161,2.870]	[2.19,2.893]	[2.139,2.820]	[2.349,2.611]
0.4	[2.310,2.922]	[2.230,2.845]	[2.260,2.879]	[2.204,2.788]	[2.384,2.609]
0.5	[2.378,2.888]	[2.298,2.808]	[2.328,2.833]	[2.269,2.755]	[2.418,2.606]
0.6	[2.446,2.854]	[2.363,2.778]	[2.39,2.795]	[2.334,2.723]	[2.453,2.603]
0.7	[2.514,2.820]	[2.429,2.744]	[2.455,2.769]	[2.399,2.691]	[2.488,2.601]
0.8	[2.582,2.786]	[2.489,2.706]	[2.51,2.73]	[2.464,2.658]	[2.523,2.598]
0.9	[2.650,2.752]	[2.556,2.676]	[2.582,2.7]	[2.528,2.626]	[2.558,2.596]
1	[2.718,2.718]	[2.619,2.639]	[2.641,2.661]	[2.593,2.593]	[2.593,2.593]

Table 4.2 Approximation errors of the vibration mass.

α	SNN	DNN	Max-Min Euler	Average Euler
0	[0.0601,0.1098]	[0.0207,0.0601]	[0.0934,0.1401]	[0.2054,0.4390]
0.2	[0.0658,0.1067]	[0.0241,0.0612]	[0.0996,0.1370]	[0.1394,0.3761]
0.6	[0.0798,0.1022]	[0.0322,0.0654]	[0.1121,0.1308]	[0.0074,0.2503]
0.8	[0.0791,0.0891]	[0.0328,0.0499]	[0.1183,0.1276]	[0.0586,0.1874]
1	[0.0921,0.0921]	[0.0534,0.0534]	[0.1246,0.1246]	[0.1246,0.1246]

Example 4.2 The heat treatment in welding can be modeled as [55]:

$$\frac{\mathrm{d}}{\mathrm{d}t}x(t) = 3Ax^2(t) \tag{4.101}$$

where transfer area A is uncertain given as $A = (1 + \alpha, 3 - \alpha)$, $\alpha \in [0, 1]$. So (4.101) is a FDE. If the initial condition is $x(0) = (0.5\sqrt{\alpha}, 0.2\sqrt{1 - \alpha} + 0.5)$, the static Bernstein neural network (4.68) has the form of

$$\begin{cases} \underline{x}_m(t, \alpha) = 0.5\sqrt{\alpha} \\ \qquad + t \sum_{i=0}^{N} \sum_{j=0}^{M} \lambda_i \gamma_j \underline{W}_{i,j} t_i (T - t_i)^{N-i} \alpha_j (1 - \alpha_j)^{M-j} \\ \overline{x}_m(t, \alpha) = 0.2\sqrt{1 - \alpha} + 0.5 \\ \qquad + t \sum_{i=0}^{N} \sum_{j=0}^{M} \lambda_i \gamma_j \overline{W}_{i,j} t_i (T - t_i)^{N-i} \alpha_j (1 - \alpha_j)^{M-j} \end{cases}. \tag{4.102}$$

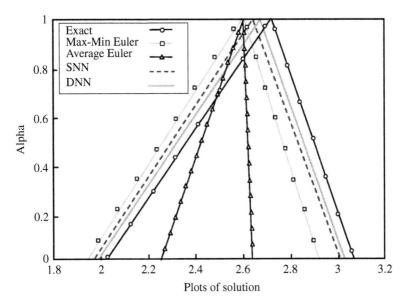

Figure 4.3 Comparison plot of the SNN, DNN, max-min Euler, and average Euler methods, and the exact solution.

With the learning rates $\eta = 0.002$ and $\gamma = 0.002$, the approximation results are shown in Table 4.3. ∎

Example 4.3 A tank system is shown in Figure 4.4. Assume $I = t + 1$ to be inflow disturbances of the tank, which generates vibration in liquid level x, where $R = 1$ is the flow obstruction that can be curbed using the valve. $A = 1$ is the cross section of the tank. The liquid level can be described as [177],

$$\frac{\mathrm{d}}{\mathrm{d}t}x(t) = -\frac{1}{AR}x(t) + \frac{I}{A} \tag{4.103}$$

The initial condition is $x(0) = (0.96 + 0.04\alpha, 1.01 - 0.01\alpha)$. The static Bernstein neural network (4.68) has the form of

$$\begin{cases} \underline{x}_m(t, \alpha) = (0.96 + 0.04\alpha) \\ \qquad + t \sum\limits_{i=0}^{N} \sum\limits_{j=0}^{M} \lambda_i \gamma_j \underline{W}_{i,j} t_i (T - t_i)^{N-i} \alpha_j (1 - \alpha_j)^{M-j} \\ \overline{x}_m(t, \alpha) = (1.01 - 0.01\alpha) \\ \qquad + t \sum\limits_{i=0}^{N} \sum\limits_{j=0}^{M} \lambda_i \gamma_j \overline{W}_{i,j} t_i (T - t_i)^{N-i} \alpha_j (1 - \alpha_j)^{M-j} \end{cases} \tag{4.104}$$

where $t \in [0, 1]$. A dynamic Bernstein neural network (4.79) is also used to approximate the solutions. To compare the results, the other generalization of the neural network method [67] is used. The comparison results are shown in

Table 4.3 Approximation errors of Bernstein neural network.

α	SNN	DNN
0	[0.0511,0.0754]	[0.0224,0.0381]
0.1	[0.0402,0.0623]	[0.0203,0.0362]
0.2	[0.0398,0.0588]	[0.0197,0.0374]
0.3	[0.0224,0.0312]	[0.0211,0.0321]
0.4	[0.0433,0.0613]	[0.0246,0.0462]
0.5	[0.0507,0.0631]	[0.0152,0.0258]
0.6	[0.0469,0.0726]	[0.0191,0.0392]
0.7	[0.0571,0.0778]	[0.0288,0.0452]
0.8	[0.0373,0.0509]	[0.0157,0.0362]
0.9	[0.0401,0.0635]	[0.0202,0.0408]
1	[0.0394,0.0394]	[0.0167,0.0167]

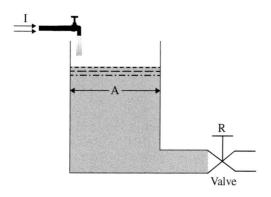

Figure 4.4 Liquid tank system.

Table 4.4. The specifications mentioned here are $\eta = 0.001$ and $\gamma = 0.001$. The corresponding errors are shown in Figure 4.5. ∎

Example 4.4 A tank with a heating system is shown in Figure 4.6, where $R = 0.5$ and the thermal capacitance is considered to be $C = 2$. The temperature is x. The model is [157],

$$\frac{\mathrm{d}}{\mathrm{d}t}x(t) = -\frac{1}{RC}x(t) \tag{4.105}$$

where $t \in [0, 1]$ and x is the amount of sinking in each moment. If the initial position is $u(0) = (\alpha - 1, 1 - \alpha)$ and $\alpha \in [0, 1]$, then the exact solutions of the fuzzy differential equation (4.105) are

$$x(t, \alpha) = [(\alpha - 1)e^t, (1 - \alpha)e^t]. \tag{4.106}$$

Table 4.4 Solutions of different methods.

α	SNN	DNN	Neural network
0	[0.0387, 0.0884]	[0.0101, 0.0398]	[0.0701, 0.1012]
0.2	[0.0451, 0.0841]	[0.0225, 0.0575]	[0.0771, 0.1207]
0.8	[0.0544, 0.0635]	[0.0144, 0.0289]	[0.0649, 0.0812]
1	[0.0554, 0.0554]	[0.0311, 0.0311]	[0.0901, 0.0901]

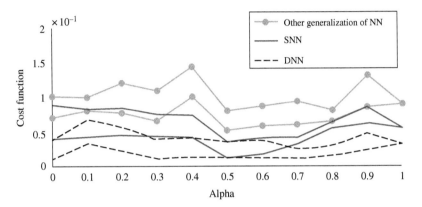

Figure 4.5 Errors between the exact solution and the approximations.

The static Bernstein neural network (4.68), is used to approximate the solution (4.106)

$$
\begin{cases}
\underline{x}_m(t, \alpha) = (\alpha - 1) \\
\quad + t \sum_{i=0}^{N} \sum_{j=0}^{M} \lambda_i \gamma_j \underline{W}_{i,j} t_i (T - t_i)^{N-i} \alpha_j (1 - \alpha_j)^{M-j} \\
\overline{x}_m(t, \alpha) = (1 - \alpha) \\
\quad + t \sum_{i=0}^{N} \sum_{j=0}^{M} \lambda_i \gamma_j \overline{W}_{i,j} t_i (T - t_i)^{N-i} \alpha_j (1 - \alpha_j)^{M-j}
\end{cases}
\tag{4.107}
$$

where $\eta = 0.001$ and $\gamma = 0.001$. A dynamic Bernstein neural network (4.79) is also used to approximate the solutions. The errors related to SNN and DNN are illustrated in Table 4.5.

For different number of learning steps $\tau = 100$, $\tau = 200$, and $\tau = 300$, and hidden neurons $n = 10$, $n = 15$, and $n = 20$, the results are shown in Table 4.6 and Figure 4.7. ∎

Both static neural network and dynamic neural network are suitable for solving the fuzzy differential equation. The leaning process of the dynamic

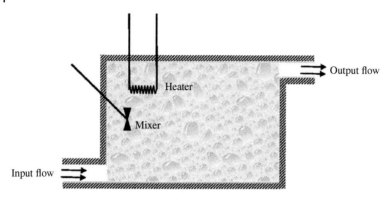

Figure 4.6 Thermal system.

Table 4.5 NN approximation errors.

α	SNN	DNN
0	[0.0407, 0.0604]	[0.0184, 0.0317]
0.1	[0.0351, 0.0578]	[0.0151, 0.0305]
0.2	[0.0334, 0.0523]	[0.0111, 0.0284]
0.8	[0.0282, 0.0417]	[0.0104, 0.0301]
0.9	[0.0253, 0.0501]	[0.0102, 0.0313]
1	[0.0323, 0.0323]	[0.0112, 0.0112]

Bernstein neural network (4.79) is faster than the static Bernstein neural network (4.68). However, the robustness of (4.68) is better than (4.79), because the weights of the dynamic Bernstein neural network are difficult to converge.

Example 4.5 A tank with a heating system is displayed in Figure 4.6, where $R = 0.5$, the thermal capacitance is $C = 2$ also the temperature is x. The model is formulated as (4.105). By utilizing the FST method, the following results are obtained

$$\mathbf{S}\left[\frac{\mathrm{d}}{\mathrm{d}t}x(t)\right] = \mathbf{S}[-x(t)] \tag{4.108}$$

$$\mathbf{S}\left[\frac{\mathrm{d}}{\mathrm{d}t}x(t)\right] = \int_0^\infty \frac{\mathrm{d}}{\mathrm{d}t}x(Bt) \odot e^{-t}\mathrm{d}t \tag{4.109}$$

where $0 \leq B < K$. If $x(t)$ is (i)-differentiable and case 1 holds, so

$$\mathbf{S}\left[\frac{\mathrm{d}}{\mathrm{d}t}x(t)\right] = \frac{1}{B} \odot (\mathbf{S}[x(t)] \ominus x(0)) = \frac{1}{B}\mathbf{S}[x(t)] \ominus \frac{1}{B}x(0). \tag{4.110}$$

Table 4.6 Different NNs.

τ	$\alpha = 0.2, n = 10$	$\alpha = 0.2, n = 15$	$\alpha = 0.2, n = 20$
100	[0.0687, 0.1087]	[0.0585, 0.0901]	[0.0487, 0.0831]
300	[0.0404, 0.0814]	[0.0392, 0.0789]	[0.0334, 0.0613]
τ	$\alpha = 0.5, n = 10$	$\alpha = 0.5, n = 15$	$\alpha = 0.5, n = 20$
100	[0.0545, 0.0957]	[0.0416, 0.0852]	[0.0352, 0.0683]
300	[0.0390, 0.0611]	[0.0291, 0.0581]	[0.0267, 0.0411]
τ	$\alpha = 0.8, n = 10$	$\alpha = 0.8, n = 15$	$\alpha = 0.8, n = 20$
100	[0.0389, 0.0855]	[0.0311, 0.0748]	[0.0219, 0.0533]
300	[0.0308, 0.0552]	[0.0206, 0.0498]	[0.0192, 0.0317]

Therefore

$$-\mathbf{S}[x(t)] = \frac{1}{B}\mathbf{S}[x(t)] \ominus \frac{1}{B}x(0). \tag{4.111}$$

Based on the Equation (4.86), the following is obtained

$$\begin{cases} -\mathbf{S}[\overline{x}(t,\alpha)] = \frac{1}{B}\mathbf{S}[\underline{x}(t,\alpha)] - \frac{1}{B}\underline{x}(0,\alpha) \\ -\mathbf{S}[\underline{x}(t,\alpha)] = \frac{1}{B}\mathbf{S}[\overline{x}(t,\alpha)] - \frac{1}{B}\overline{x}(0,\alpha) \end{cases}. \tag{4.112}$$

Therefore, the solution of Equation (4.112) is as follows

$$\begin{cases} \mathbf{S}[\overline{x}(t,\alpha)] = \left(\frac{-1}{B^2-1}\right)\overline{x}(0,\alpha) + \left(\frac{B}{B^2-1}\right)\underline{x}(0,\alpha) \\ \mathbf{S}[\underline{x}(t,\alpha)] = \left(\frac{-1}{B^2-1}\right)\underline{x}(t,\alpha) + \left(\frac{B}{B^2-1}\right)\overline{x}(0,\alpha) \end{cases}. \tag{4.113}$$

By utilizing the inverse Sumudu transform, the following is obtained

$$\begin{cases} \mathbf{S}[\overline{x}(t,\alpha)] = \overline{x}(0,\alpha)\mathbf{S}^{-1}\left(\frac{-1}{B^2-1}\right) + \underline{x}(0,\alpha)\mathbf{S}^{-1}\left(\frac{B}{B^2-1}\right) \\ \mathbf{S}[\underline{x}(t,\alpha)] = \underline{x}(0,\alpha)\mathbf{S}^{-1}\left(\frac{-1}{B^2-1}\right) + \overline{x}(0,\alpha)\mathbf{S}^{-1}\left(\frac{B}{B^2-1}\right) \end{cases} \tag{4.114}$$

where

$$\begin{cases} \overline{x}(t,\alpha) = e^t\left(\frac{\overline{x}(0,\alpha)-\underline{x}(0,\alpha)}{2}\right) + e^{-t}\left(\frac{\overline{x}(0,\alpha)+\underline{x}(0,\alpha)}{2}\right) \\ \underline{x}(t,\alpha) = e^t\left(\frac{\underline{x}(0,\alpha)-\overline{x}(0,\alpha)}{2}\right) + e^{-t}\left(\frac{\underline{x}(0,\alpha)+\overline{x}(0,\alpha)}{2}\right). \end{cases} \tag{4.115}$$

Now if $x(t)$ be (ii)-differentiable and case 2 holds, then

$$\mathbf{S}\left[\frac{\mathrm{d}}{\mathrm{d}t}x(t)\right] = \left(\frac{-1}{B}\mathbf{S}[x(t)]\right) \ominus \left(\frac{-1}{B}x(0)\right). \tag{4.116}$$

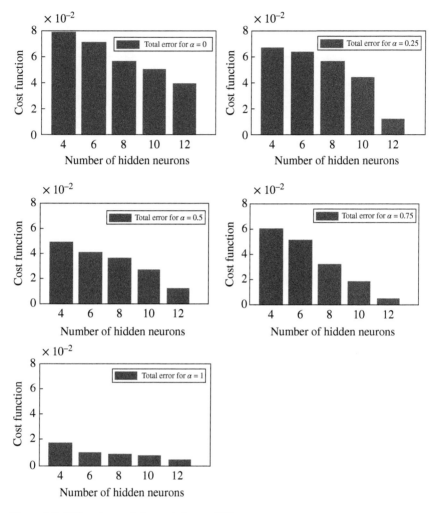

Figure 4.7 Different neural elements for $\tau = 300$.

Hence

$$-\mathbf{S}[x(t)] = \left(\frac{-1}{B}\mathbf{S}[x(t)]\right) \ominus \left(\frac{-1}{B}x(0)\right). \tag{4.117}$$

Based on the above relations, Equation (4.105) can be written as follows

$$\begin{cases} -\mathbf{S}[\underline{x}(t,\alpha)] = \frac{1}{B}\mathbf{S}[\underline{x}(t,\alpha)] - \frac{1}{B}\underline{x}(0,\alpha) \\ -\mathbf{S}[\overline{x}(t,\alpha)] = \frac{1}{B}\mathbf{S}[\overline{x}(t,\alpha)] - \frac{1}{B}\overline{x}(0,\alpha) \end{cases}. \tag{4.118}$$

So, the solution of Eq. (4.118) is displayed as

$$
\begin{cases}
S[\underline{x}(t,\alpha)] = \underline{x}(0,\alpha)\left(\frac{1}{B+1}\right) \\
S[\overline{x}(t,\alpha)] = \overline{x}(t,\alpha)\left(\frac{1}{B+1}\right).
\end{cases}
\tag{4.119}
$$

By utilizing the inverse Sumudu transform, the following is obtained

$$
\begin{cases}
\underline{x}(t,\alpha) = \underline{x}(0,\alpha)S^{-1}\left(\frac{1}{B+1}\right) \\
\overline{x}(t,\alpha) = \overline{x}(0,\alpha)S^{-1}\left(\frac{1}{B+1}\right)
\end{cases}
\tag{4.120}
$$

where

$$
\begin{cases}
\underline{x}(t,\alpha) = e^{-t}\underline{x}(0,\alpha) \\
\overline{x}(t,\alpha) = e^{-t}\overline{x}(0,\alpha)
\end{cases}.
\tag{4.121}
$$

If the initial condition be a symmetric triangular fuzzy number as $x(0) = (-a(1-\alpha), a(1-\alpha))$, then the following cases will be hold.
Case 1 :

$$
\begin{cases}
\underline{x}(t,\alpha) = e^{t}(-a(1-\alpha)) \\
\overline{x}(t,\alpha) = e^{t}(a(1-\alpha))
\end{cases}.
\tag{4.122}
$$

Case 2:

$$
\begin{cases}
\underline{x}(t,\alpha) = e^{-t}(-a(1-\alpha)) \\
\overline{x}(t,\alpha) = e^{-t}(a(1-\alpha))
\end{cases}
\tag{4.123}
$$

Corresponding solution plots are displayed in Figure 4.8 and Figure 4.9. Corresponding error plots are shown in Figure 4.10. These errors are the differences of the exact and the approximation solutions, for two different methods: FST and Average Euler method [185]. FST is more accurate than the Average Euler method. ∎

Example 4.6 A tank system is shown in Figure 4.4. Suppose $I = t + 1$ is inflow disturbances of the tank that generates vibration in liquid level ϕ, also $H = 1$ is the flow obstruction, which can be curbed utilizing the valve. $Q = 1$ is the cross section of the tank. The liquid level is illustrated as (4.103). By utilizing the FST method the following is obtained

$$
-S[x(t)] = \left(\frac{1}{B} \odot S[x(t)]\right) \ominus \left(\frac{1}{B}S[x(0)]\right)
\tag{4.124}
$$

$$
S\left[\frac{d}{dt}x(t)\right]\left(\int_{a_1}^{\infty}\frac{d}{dt}x(Bt)e^{-t}dt\right)
\tag{4.125}
$$

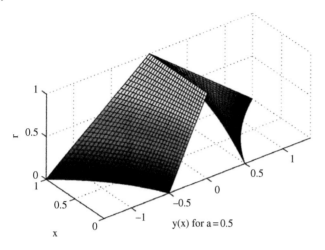

Figure 4.8 Solution of the FDE under case 1 consideration.

Figure 4.9 Solution of FDE under case 2 consideration.

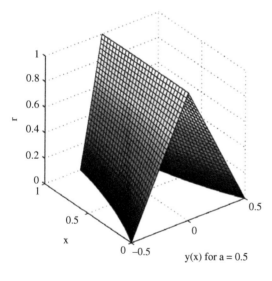

For this example, case 2 is only considered, since the procedure of case 1 is same as that of case 2. By considering case 2 the following relation is obtained

$$\mathbf{S}\left[\frac{d}{dt}x(t)\right] = \left(\frac{-1}{B} \odot \mathbf{S}[x(t)]\right) \ominus \left(\frac{-1}{B}\mathbf{S}[x(0)]\right). \tag{4.126}$$

So

$$-\mathbf{S}[x(t)] + \mathbf{S}[t] + \mathbf{S}[1] = \left(\frac{-1}{B} \odot \mathbf{S}[x(t)]\right) \ominus \left(\frac{-1}{B}\mathbf{S}[x(0)]\right). \tag{4.127}$$

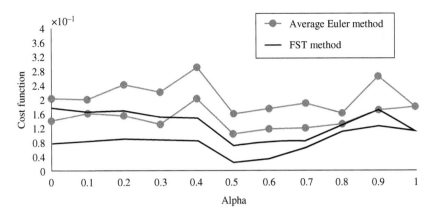

Figure 4.10 Lower and upper bounds of absolute errors.

Based on the Equation (4.86), the following is obtained

$$\begin{cases} -S[\underline{x}(t,\alpha)] + S[t] + S[1] = \frac{1}{B}S[\underline{x}(t,\alpha)] - \frac{1}{B}\underline{x}(0,\alpha) \\ -S[\overline{x}(t,\alpha)] + S[t] + S[1] = \frac{1}{B}S[\overline{x}(t,\alpha)] - \frac{1}{B}\overline{x}(0,\alpha) \end{cases} \quad (4.128)$$

Therefore, the solution of Equation (4.128) is extracted as

$$\begin{cases} S[\underline{x}(t,\alpha)] = S[t] + S[1] + \frac{-1}{B}S[\underline{x}(t,\alpha)] - \frac{1}{B}\underline{x}(0,\alpha) \\ S[\overline{x}(t,\alpha)] = S[t] + S[1] + \frac{1}{B}S[\overline{x}(t,\alpha)] - \frac{1}{B}\overline{x}(0,\alpha) \end{cases} \quad (4.129)$$

Hence,

$$\begin{cases} S[\underline{x}(t,\alpha)] = \left(\frac{1}{B+1}\right)\underline{x}(0,\alpha) + B \\ S[\overline{x}(t,\alpha)] = \left(\frac{1}{B+1}\right)\overline{x}(t,\alpha) + B \end{cases} \quad (4.130)$$

By utilizing the inverse Sumudu transform, the following is obtained

$$\begin{cases} \underline{x}(t,\alpha) = \underline{x}(0,\alpha)S^{-1}\left(\frac{1}{B+1}\right) + S^{-1}(B) \\ \overline{x}(t,\alpha) = \overline{x}(0,\alpha)S^{-1}\left(\frac{1}{B+1}\right) + S^{-1}(B) \end{cases} \quad (4.131)$$

where

$$\begin{cases} \underline{x}(t,\alpha) = e^{-t}\underline{x}(0,\alpha) + t \\ \overline{x}(t,\alpha) = e^{-t}\overline{x}(0,\alpha) + t \end{cases} \quad (4.132)$$

If the initial condition is taken to be a symmetric triangular fuzzy number as $x(0) = (-a(1-\alpha), a(1-\alpha))$, so

$$\begin{cases} \underline{x}(t,\alpha) = e^{-t}(-a(1-\alpha)) + t \\ \overline{x}(t,\alpha) = e^{-t}(a(1-\alpha)) + t \end{cases} \quad (4.133)$$

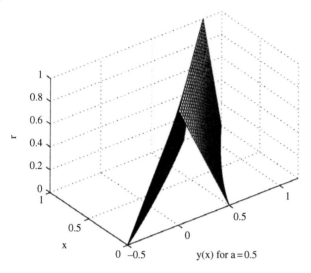

Figure 4.11 Solution of the FDE under case 2 consideration.

Figure 4.11 shows the Corresponding solution plot. Figure 4.12 shows the corresponding error plots. These errors are the differences of the exact and the approximation solutions, for two different methods: FST and Max-Min Euler method [185]. It can be seen that both the FST method and the Max-Min Euler method can approximate the solutions of the fuzzy initial value problem (4.103). The approximation errors of the FST method are much smaller than the Max-Min Euler method. ∎

Example 4.7 A nuclear decay equation can be described as [40],

$$\left\{ \frac{\mathrm{d}}{\mathrm{d}t} N(t) = -\lambda N(t) \right. \tag{4.134}$$

where $N(t)$ is considered to be the number of radionuclides present, λ is state as the decay constant, also N_0 is taken to be the initial number of radionuclides. Let N_0 be a fuzzy number. By utilizing the FST method the following outcomes can be demonstrated

$$\mathbf{S}\left[\frac{\mathrm{d}}{\mathrm{d}t} N(t) \right] = \mathbf{S}[-\lambda N(t)] = -\lambda \mathbf{S}[N(t)] \tag{4.135}$$

$$\mathbf{S}\left[\frac{\mathrm{d}}{\mathrm{d}t} N(t) \right] = \int_a^\infty \frac{\mathrm{d}}{\mathrm{d}t} N(st) e^{-t} \mathrm{d}t. \tag{4.136}$$

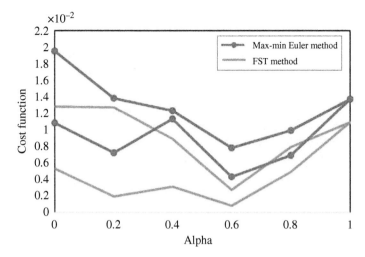

Figure 4.12 Lower and upper bounds of absolute errors.

For this example, case 2 is only considered, since the procedure of case 1 is same as that of case 2. If $N(t)$ is (ii)-differentiable and case 2 holds, then

$$\mathbf{S}\left[\frac{\mathrm{d}}{\mathrm{d}t}N(t)\right] = \left(\frac{-1}{B}\mathbf{S}[N(t)]\right) \ominus \left(\frac{-1}{B}N(0)\right). \tag{4.137}$$

Therefore

$$-\lambda\mathbf{S}[N(t)] = \left(\frac{-1}{B}\mathbf{S}[N(t)]\right) \ominus \left(\frac{-1}{B}N(0)\right). \tag{4.138}$$

According to Equation (4.86), the relation below is extracted

$$\begin{cases} -\lambda\mathbf{S}[\underline{N}(t,\alpha)] = \frac{1}{B}\mathbf{S}[\underline{N}(t,\alpha)] - \frac{1}{B}\underline{N}(0,\alpha) \\ -\lambda\mathbf{S}[\overline{N}(t,\alpha)] = \frac{1}{B}\mathbf{S}[\overline{N}(t,\alpha)] - \frac{1}{B}\overline{N}(0,\alpha) \end{cases}. \tag{4.139}$$

Hence, the solution of Equation (4.139) is as follows

$$\begin{cases} \mathbf{S}[\underline{N}(t,\alpha)]\left(\lambda - \frac{1}{B}\right) = \frac{1}{B}\underline{N}(0,\alpha) \\ \mathbf{S}[\overline{N}(t,\alpha)]\left(\lambda - \frac{1}{B}\right) = \frac{1}{B}\overline{N}(t,\alpha) \end{cases}. \tag{4.140}$$

Thus,

$$\begin{cases} \mathbf{S}[\underline{N}(t,\alpha)] = \frac{1}{-\lambda B+1}\underline{N}(0,\alpha) \\ \mathbf{S}[\overline{N}(t,\alpha)] = \frac{1}{-\lambda B+1}\overline{N}(t,\alpha) \end{cases}. \tag{4.141}$$

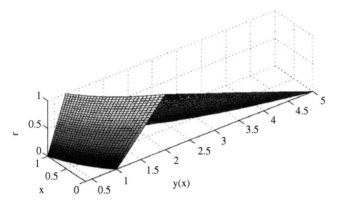

Figure 4.13 Solution of the FDE under case 2 consideration.

So, by utilizing the inverse Sumudu transform, the following are obtained

$$
\begin{cases}
\underline{N}(t, \alpha) = \underline{N}(0, \alpha)\mathbf{S}^{-1}\left(\frac{1}{-\lambda B+1}\right) \\
\overline{N}(t, \alpha) = \overline{N}(0, \alpha)\mathbf{S}^{-1}\left(\frac{1}{-\lambda B+1}\right)
\end{cases}
\tag{4.142}
$$

where

$$
\begin{cases}
\underline{N}(t, \alpha) = e^{-\lambda t}\underline{N}(0, \alpha) \\
\overline{N}(t, \alpha) = e^{-\lambda t}\overline{N}(0, \alpha)
\end{cases}.
\tag{4.143}
$$

Let $\lambda = 1$ and $N_0 = (1, 2, 5)$, then

$$
\begin{cases}
\underline{N}(0, \alpha) = (1 + \alpha) \\
\overline{N}(0, \alpha) = (5 - 3\alpha)
\end{cases}.
\tag{4.144}
$$

So

$$
\begin{cases}
\underline{N}(t, \alpha) = e^{-t}(1 + \alpha) \\
\overline{N}(t, \alpha) = e^{-t}(5 - 3\alpha)
\end{cases}.
\tag{4.145}
$$

The corresponding solution plot is displayed in Figure 4.13. Figure 4.14 shows the corresponding error plots. These errors are the differences of the exact and the approximation solutions, for two different methods: FST and Euler method [121]. It can be seen that both the FST method and the Euler method can approximate the solutions of the fuzzy initial value problem (4.134). The approximation errors of the FST method are much smaller than the Euler method. ∎

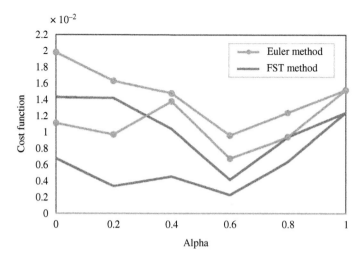

Figure 4.14 Lower and upper bounds of absolute error.

4.7 Summary

In this chapter, the solutions of the FDEs are approximated by two types of Bernstein neural networks. Initially, the FDE is transformed into four ODEs with Hukuhara differentiability. Then neural models are constructed with the structure of ODEs. A modified back propagation method for fuzzy variables is used for training the neural networks. The theoretical analysis and simulation results show that these new models, Bernstein neural networks, are effective to estimate the solutions of FDEs.

Furthermore, a new method based on FST is used to obtain the approximate solutions of FDEs. Significant theorems are suggested in order to explain the properties of FST.

5

System Modeling with Partial Differential Equations

5.1 Introduction

In this chapter, for solving the strongly degenerate parabolic and Burgers–Fisher equations, a model using neural networks is developed. A trial solution of this system is subdivided into two parts. The initial and boundary conditions comprise the first part, which contains no adjustable parameters. The involvement of a neural network containing adjustable parameters (weights and biases) comprise the second part. A training technique is implemented for the training of the network, which requires the calculation of the error gradient in consideration of the network parameters. In order to maintain a continuation, a significantly modified methodology is illustrated for solving a wave equation that is modeled on the basis of two patterned Bernstein neural networks: static and dynamic models.

Some numerical examples are proposed to show the effectiveness of the approximation methods with neural networks and Bernstein neural networks.

5.2 Solutions using Burgers–Fisher Equations

Definition 5.1 (Second order nonlinear partial differential equation (PDE)) *The second order singular nonlinear PDE can be illustrated using the equation below*

$$\frac{\partial^2 u(x,t)}{\partial t^2} + \frac{2}{t}\frac{\partial u(x,t)}{\partial t} = F\left(x, u(x,t), \frac{\partial u(x,t)}{\partial x}, \frac{\partial^2 u(x,t)}{\partial x^2}\right) \tag{5.1}$$

where t and x are independent variables, u is a dependent variable, F is a non-linear function of x, u, u_x, and u_{xx}, and the initial conditions for the PDE (5.1) are as follows

$$u(x,0) = f(x), \quad u_t(x,0) = g(x). \tag{5.2}$$

Modeling and Control of Uncertain Nonlinear Systems with Fuzzy Equations and Z-Number,
First Edition. Wen Yu and Raheleh Jafari.
© 2019 by The Institute of Electrical and Electronics Engineers, Inc. Published 2019 by John Wiley & Sons, Inc.

Definition 5.2 (Strongly degenerate parabolic equation) *The strongly degenerate parabolic equation is explained as follows*

$$\frac{\partial u(x,t)}{\partial t} + \frac{\partial Q(u(x,t))}{\partial x} = \frac{\partial^2 A(u(x,t))}{\partial x^2},$$

$$(x,t) \in \Pi_T := [0,1] \times (0,T), \quad T > 0. \tag{5.3}$$

Consider the boundary conditions as

$$u(x,0) = g_0(x), \quad u(0,t) = f_0(t), \quad u(1,t) = f_1(t) \tag{5.4}$$

where the integrated diffusion coefficient A is exhibited by

$$A(u) = \int_0^u a(s)\mathrm{d}s, \quad a(u) \geq 0, \quad a \in L^\infty([0,1]) \cap L^1([0,1]). \tag{5.5}$$

The function a is allowed to terminate on u intervals of positive length, in which Equation (5.3) degenerates to a first order scalar conservation law. Hence, Equation (5.3) is termed strongly degenerate.

Definition 5.3 (Burgers–Fisher equation) *The generalized Burgers–Fisher equation is defined as*

$$\frac{\partial u(x,t)}{\partial t} + \alpha u^\sigma(x,t)\frac{\partial u(x,t)}{\partial x} - \frac{\partial^2 u(x,t)}{\partial x^2} = \beta u(x,t)(1 - u^\sigma(x,t))$$

$$(x,t) \in \Pi_T := [0,1] \times [0,T], \quad T > 0 \tag{5.6}$$

with initial and boundary conditions

$$g_0(x) := u(x,0) = \left(\frac{1}{2} + \frac{1}{2}\tanh\left(\frac{-\alpha\sigma}{2(\sigma+1)}x\right)\right)^{\frac{1}{\sigma}}, \quad 0 \leq x \leq 1 \tag{5.7}$$

$$f_0(t) := u(0,t) = \left(\frac{1}{2} + \frac{1}{2}\tanh\left(\frac{-\alpha\sigma}{2(\sigma+1)}\left(-\left(\frac{\alpha}{\sigma+1} + \frac{\beta(\sigma+1)}{\alpha}\right)t\right)\right)\right)^{\frac{1}{\sigma}},$$

$$0 \leq x \leq 1, \quad t \geq 0 \tag{5.8}$$

$$f_1(t) := u(1,t) = \left(\frac{1}{2} + \frac{1}{2}\tanh\left(\frac{-\alpha\sigma}{2(\sigma+1)}\left(1 - \left(\frac{\alpha}{\sigma+1} + \frac{\beta(\sigma+1)}{\alpha}\right)t\right)\right)\right)^{\frac{1}{\sigma}},$$

$$0 \leq x \leq 1, \quad t \geq 0 \tag{5.9}$$

where α, β, and σ are constants.

Here a three layer neural network with two input signals, m hidden neurons, and one output signal is formulated in order to solve the strongly degenerate parabolic and Burgers–Fisher equations, which depend on the function approximation ability of the neural network and repays the solution

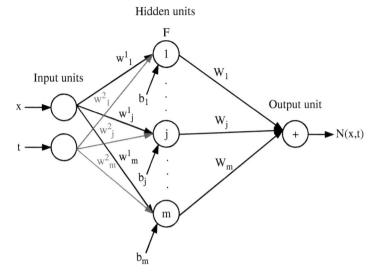

Figure 5.1 Neural network equivalent to strongly degenerate parabolic and Burgers–Fisher equations.

of differential equations in a closed analytic and differentiable form (see Figure 5.1). The input–output connection of each unit of the suggested neural network is written as

- Input units:

$$o_1^1 = x$$
$$o_2^1 = t. \tag{5.10}$$

- Hidden units:

$$o_j^2 = F(b_j + w_j^1 x + w_j^2 t) \quad j = 1, \dots, m. \tag{5.11}$$

- Output units:

$$N(x, t) = \sum_{j=1}^{m} (W_j o_j^2). \tag{5.12}$$

Here, the most common function $F(r) = \frac{2}{1+e^{-2r}} - 1$ (tan-sigmoid function), which is a continuously differentiable nonlinear function, is accepted as an activation function of the hidden units. In order to set the boundary conditions, a trial solution is selected, which is the sum of two terms, as

$$
\begin{aligned}
u_m(x, t) = (1 - x)f_0(t) &+ xf_1(t) + (1 - t)\{g_0(x) \\
&- [(1 - x)g_0(0) + xg_0(1)]\} + x(1 - x)tN(x, t)
\end{aligned} \tag{5.13}
$$

where

$$N(x,t) = \sum_{j=1}^{m}(W_j F(b_j + w_j^1 x + w_j^2 t)). \tag{5.14}$$

If it is supposed that $0 \leq x \leq 1$, the rectangle $[0,1] \times [0,T]$ can be divided into nn' mesh points $(x_i, t_j) = ((i-1)h, (j-1)h')$, $h = \frac{1}{n-1}$, $h' = \frac{T}{n'-1}$, $(i = 1, \ldots, n; j = 1, \ldots, n')$. In the given problems, the substitution of approximate solution $u_m(x,t)$ instead of the unknown function will result in the following least mean square error for the relation mentioned, $(x,t) = (x_i, t_j)$, as follows

$$E_{i,j} = \frac{1}{2}(M_{i,j})^2 \tag{5.15}$$

where for the Burgers–Fisher equation

$$M_{i,j} = \frac{\partial u_m(x,t)}{\partial t}\Big|_{\substack{x=x_i \\ t=t_j}} + \alpha u_m^\sigma(x,t)\Big|_{\substack{x=x_i \\ t=t_j}} \frac{\partial u_m(x,t)}{\partial x}\Big|_{\substack{x=x_i \\ t=t_j}} - \frac{\partial^2 u_m(x,t)}{\partial x^2}\Big|_{\substack{x=x_i \\ t=t_j}}$$

$$-\beta u_m(x,t)\Big|_{\substack{x=x_i \\ t=t_j}} (1 - u_m^\sigma(x,t)\Big|_{\substack{x=x_i \\ t=t_j}}) \tag{5.16}$$

and for a strongly degenerate parabolic equation

$$M_{i,j} = \frac{\partial u_m(x,t)}{\partial t}\Big|_{\substack{x=x_i \\ t=t_j}} + \frac{\partial Q(u_m(x,t))}{\partial x}\Big|_{x=x_i, t=t_j} - \frac{\partial^2 A(u_m(x,t))}{\partial x^2}\Big|_{\substack{x=x_i \\ t=t_j}}.$$

$$\tag{5.17}$$

Generally the summed up error of the suggested neural network is illustrated as

$$E = \sum_{i=1}^{n}\sum_{j=1}^{n'} E_{i,j} = \sum_{i,j} E_{i,j} \tag{5.18}$$

The proposed learning rule can be systematically obtained as minimizers of the referred error function. The above statement indicates that the main intention in the remaining part of this section is to train the proposed network architecture in order to complete this task. Consideration of the above intention will start the process of derivation of a learning procedure that is considered to be a natural generalization of the Newton method in order to adjust network parameters (weights and biases). The Newton rule is performed on the basis of cost function $E_{i,j}$, to calculate the weight modifications W_q as [150]

$$W_q(r+1) = W_q(r) - \mu(r)\frac{\partial E_{i,j}}{\partial W_q}, \quad q = 1, \ldots, m \tag{5.19}$$

where μ is the training rate, $\mu > 0$. In order to speed up the training process, a momentum term is added as

$$W_q(r+1) = W_q(r) - \mu(r)\frac{\partial E_{i,j}}{\partial W_q} + \gamma[W_q(r) - W_q(r-1)] \tag{5.20}$$

where $\gamma > 0$. The index r refers to the iteration number. In addition, $W_q(r+1)$ and $W_q(r)$ are the updated and recent output weight values, respectively. Now, the proposed methodology of (5.19) can be illustrated as follows

$$\begin{bmatrix} W_1 \\ \vdots \\ W_m \end{bmatrix}_{r+1} = \begin{bmatrix} W_1 \\ \vdots \\ W_m \end{bmatrix}_r - \frac{(\nabla E_{i,j}(W)_r)^T \nabla E_{i,j}(W)_r}{(\nabla E_{i,j}(W)_r)^T Q_r \nabla E_{i,j}(W)_r} \nabla E_{i,j}(W)_r + \gamma \begin{bmatrix} \Delta W_1 \\ \vdots \\ \Delta W_m \end{bmatrix}_{r-1} \tag{5.21}$$

where

$$\nabla E_{i,j}(W) = \left(\frac{\partial E_{i,j}}{\partial W_1}, \cdots, \frac{\partial E_{i,j}}{\partial W_m} \right)^T \tag{5.22}$$

and

$$Q = \begin{bmatrix} \dfrac{\partial^2 E_{i,j}}{\partial W_1^2} & \dfrac{\partial^2 E_{i,j}}{\partial W_1 \partial W_2} & \cdots & \dfrac{\partial^2 E_{i,j}}{\partial W_1 \partial W_m} \\ \dfrac{\partial^2 E_{i,j}}{\partial W_2 \partial W_1} & \dfrac{\partial^2 E_{i,j}}{\partial W_2^2} & \cdots & \dfrac{\partial^2 E_{i,j}}{\partial W_2 \partial W_m} \\ \cdots & \cdots & \cdots & \cdots \\ \dfrac{\partial^2 E_{i,j}}{\partial W_m \partial W_1} & \dfrac{\partial^2 E_{i,j}}{\partial W_m \partial W_2} & \cdots & \dfrac{\partial^2 E_{i,j}}{\partial W_m^2} \end{bmatrix} \tag{5.23}$$

are computed at the current mesh points (x_i, t_j). Here Q is the Hessian matrix with components $\frac{\partial^2 E_{i,j}}{\partial W_q \partial W_p}$ (for $q, p = 1, \ldots, m$). It is quite obvious that the convergence speed has a direct relation to the learning rate. In order to achieve an optimal learning rate for rapid convergence of the learning optimization rule, the inverse of the Hessian matrix Q of the error function $E_{i,j}$ is implemented at the current mesh points. The approximate Newton method illustrated above is sufficient to scale the descent in each step. The usage of the chain rule for differentiation will result in the present partial derivative, as mentioned below

$$\frac{\partial E_{i,j}}{\partial W_q} = \frac{\partial E_{i,j}}{\partial M_{i,j}} \cdot \frac{\partial M_{i,j}}{\partial W_q} = M_{i,j} \cdot \frac{\partial M_{i,j}}{\partial W_q}. \tag{5.24}$$

For the Burgers–Fisher equation assume that

$$K_1(x,t) = \frac{\partial u(x,t)}{\partial t}, \quad K_2(x,t) = u^\sigma(x,t)\frac{\partial u(x,t)}{\partial x}$$

$$K_3(x,t) = \frac{\partial^2 u(x,t)}{\partial x^2}, \quad K_4(x,t) = u(x,t)(1 - u^\sigma(x,t)) \tag{5.25}$$

and then the following is concluded

$$
\begin{aligned}
\frac{\partial M_{i,j}}{\partial W_q} = {} & \frac{\partial M_{i,j}}{\partial K_1(x_i,t_j)} \cdot \frac{\partial K_1(x_i,t_j)}{\partial W_q} + \alpha \frac{\partial M_{i,j}}{\partial K_2(x_i,t_j)} \cdot \frac{\partial K_2(x_i,t_j)}{\partial W_q} \\
& - \frac{\partial M_{i,j}}{\partial K_3(x_i,t_j)} \cdot \frac{\partial K_3(x_i,t_j)}{\partial W_q} - \beta \frac{\partial M_{i,j}}{\partial K_4(x_i,t_j)} \cdot \frac{\partial K_4(x_i,t_j)}{\partial W_q}.
\end{aligned}
\tag{5.26}
$$

For the strongly degenerate parabolic equation used in the stated methodology, $\frac{\partial M_{i,j}}{\partial W_q}$ can be obtained in the same manner. When the above relation is substituted into (5.21), the desired learning rule will be achieved. The learning procedure stated above can also be extended to the other network parameters (w_q^1, w_q^2, and b_q) in a similar manner.

5.3 Solution using Wave Equations

Definition 5.4 (Wave equation) *The Cauchy problem for the wave equation in one dimension can be stated as*

$$
\frac{\partial^2 u(x,t)}{\partial t^2} + c^2 \frac{\partial^2 u(x,t)}{\partial x^2} = f(x,t), \quad (x,t) \in [0,a] \times [0,b]
\tag{5.27}
$$

with

$$
u(x,0) = \phi(x), \quad u_t(x,0) = \psi(x)
\tag{5.28}
$$

where a and b are constants. In above equation the parameter c is the speed of light.

Here PDEs are solved with the help of two patterns of neural networks and the application of the Bernstein polynomial. Consider the Cauchy problem (5.27), where the solution u depends on both spatial and temporal variables x and t, respectively. The trial solution is written as

$$
u_m(x,t) = \phi(x) + t\psi(x) + t\left[B(x,t) - B(x,0) - \frac{\partial B(x,0)}{\partial t}\right]
\tag{5.29}
$$

where $B(x,t)$ is the bivariate Bernstein polynomial series of solution function $u(x,t)$, which is

$$
B(x,t) = \sum_{i=0}^{n}\sum_{j=0}^{m} \binom{n}{i}\binom{m}{j} \frac{x^i(a-x)^{n-i}}{a^n} \frac{t^j(b-t)^{m-j}}{b^m} q_{i,j}(x,t), \quad n,m \in N.
\tag{5.30}
$$

So

$$
B(x,t) = \sum_{i=0}^{n}\sum_{j=0}^{m} \beta_{i,j} x^i(a-x)^{n-i} t^j(b-t)^{m-j} q_{i,j}(x,t), \quad n,m \in N,
$$

$$\beta_{i,j} = \frac{1}{a^n b^m} \binom{n}{i} \binom{m}{j} \tag{5.31}$$

where

$$\binom{n}{i} = \frac{n!}{i!(n-i)!}, \quad \binom{m}{j} = \frac{m!}{j!(m-j)!}. \tag{5.32}$$

Consider the following relations

$$\frac{\partial^2 u_m(x,t)}{\partial x^2} = \phi''(x) + t\psi''(x) + t\left[\frac{\partial^2 B(x,t)}{\partial x^2} - \frac{\partial^2 B(x,0)}{\partial x^2} - \frac{\partial^2 \partial B(x,0)}{\partial x^2 \partial t}\right] \tag{5.33}$$

and

$$\frac{\partial^2 u_m(x,t)}{\partial t^2} = 2\frac{\partial B(x,t)}{\partial t} + t\frac{\partial^2 B(x,t)}{\partial t^2}. \tag{5.34}$$

Substituting the above relations in the origin problem (5.27) leads to the following differential equation

$$2\frac{\partial B(x,t)}{\partial t} + t\frac{\partial^2 B(x,t)}{\partial t^2} + c^2(\phi''(x) + t\psi''(x)$$
$$+ t\left[\frac{\partial^2 B(x,t)}{\partial x^2} - \frac{\partial^2 B(x,0)}{\partial x^2} - \frac{\partial^2 \partial B(x,0)}{\partial x^2 \partial t}\right]\right) = f(x,t)$$
$$(x,t) \in [0,a] \times [0,b]. \tag{5.35}$$

For simplicity the above relation can be rewritten as follows

$$\sum_{i=0}^{n} \sum_{j=0}^{m} \xi_{i,j}(x,t) q_{i,j}(x,t) + \zeta(x,t) = f(x,t), \quad (x,t) \in [0,a] \times [0,b] \tag{5.36}$$

where

$$\xi_{i,j}(x,t) = 2\beta_{i,j} x^i (a-x)^{n-i} (jt^{j-1}(b-t)^{m-j} - (m-j)t^j(b-t)^{m-j-1})$$
$$+ t\beta_{i,j} x^i (a-x)^{n-i} (j(j-1)t^{j-2}(b-t)^{m-j} - 2j(m-j)t^{j-1}(b-t)^{m-j-1}$$
$$+ (m-j)(m-j-1)t^j(b-t)^{m-j-2}) + c^2 t\beta_{i,j}(i(i-1)x^{i-2}(a-x)^{n-i}$$
$$- 2i(n-i)x^{i-1}(a-x)^{n-i-1} + (n-i)(n-i-1)x^i(a-x)^{n-i-2})t^j(b-t)^{m-j}$$
$$- c^2 t\beta_{i,j}(i(i-1)x^{i-2}(a-x)^{n-i} - 2i(n-i)x^{i-1}(a-x)^{n-i-1}$$
$$+ (n-i)(n-i-1)x^i(a-x)^{n-i-2})(jt^{j-1}(b-t)^{m-j} - (m-j)t^j(b-t)^{m-j-1}) \tag{5.37}$$

and

$$\zeta(x,t) = c^2 \left(\phi''(x) + t\psi''(x) + t\frac{\partial^2 B(x,0)}{\partial x^2}\right). \tag{5.38}$$

A neural network is designed to represent the equation (5.31), see Figure 4.2.

In the above architecture the mathematical symbol is defined as

$$\varphi_{i,j} = \sum_{i=0}^{n} \sum_{j=0}^{m} \binom{n}{i} \binom{m}{j} \frac{x^i (a-x)^{n-i}}{a^n} \frac{t^j (b-t)^{m-j}}{b^m}, \quad n, m \in N. \quad (5.39)$$

The input–output relation of each unit in the proposed neural architecture can be summarized as follows.

- Input unit

$$o_{i,j} = q_{i,j}, \quad i = 0, \ldots, n, \quad j = 0, \ldots, m. \quad (5.40)$$

- Output unit

$$N(x, t) = \varphi_{i,j} o_{i,j}. \quad (5.41)$$

Now, a suitable numerical technique should be able to provide an appropriate tool for measuring and calculating the accuracy of the obtained solution. Hence, to compare the exact solution with its obtained one, the least mean square error is used, which is stated as follows

$$E_{i,j} = \frac{1}{2} \left(\sum_{i=0}^{n} \sum_{j=0}^{m} \xi_{i,j}(x, t) q_{i,j}(x, t) + \zeta(x, t) - f(x, t) \right)^2. \quad (5.42)$$

Newton's rule as described in (5.21) is used for adjusting the parameters such that the network error is minimized over the space of the weights setting. The initial parameter $q_{i,j}$ is selected randomly to begin the procedure. The described standard self-learning mechanism works as follows

$$q_{i,j}(r+1) = q_{i,j}(r) - \mu(r) \frac{\partial E_{i,j}}{\partial q_{i,j}} \quad (5.43)$$

where μ is the training rate, $\mu > 0$. In order to speed up the training process, a momentum term is added as

$$q_{i,j}(r+1) = q_{i,j}(r) - \mu(r) \frac{\partial E_{i,j}}{\partial q_{i,j}} + \gamma [q_{i,j}(r) - q_{i,j}(r-1)] \quad (5.44)$$

where $\gamma > 0$. The index r refers to the iteration number.

Consider another type of neural network architecture shown in Figure 4.1. The input–output relation of each unit in the proposed neural architecture can be summarized as follows.

- Input unit

$$o_1^1 = x$$
$$o_2^1 = t. \quad (5.45)$$

- The first hidden units

$$o_{1,i}^2 = f_i^1(o_1^1), \quad o_{2,i}^2 = f_i^2(o_1^1), \quad i = 0, \dots, n$$
$$o_{3,j}^2 = g_j^1(o_2^1), \quad o_{4,j}^2 = g_j^2(o_2^1), \quad j = 0, \dots, m. \tag{5.46}$$

- The second hidden units

$$o_{1,i}^3 = o_{1,i}^2(o_{2,i}^2), \quad i = 0, \dots, n$$
$$o_{2,j}^3 = o_{3,j}^2(o_{4,j}^2), \quad j = 0, \dots, m. \tag{5.47}$$

- The third hidden units

$$o_{1,i}^4 = \lambda_i o_{1,i}^3, \quad i = 0, \dots, n$$
$$o_{2,j}^4 = \gamma_j o_{2,j}^3, \quad j = 0, \dots, m. \tag{5.48}$$

- The fourth hidden units

$$o_{i,j}^5 = o_{1,i}^4 o_{2,j}^4, \quad i = 0, \dots, n, \quad j = 0, \dots, m. \tag{5.49}$$

- Output unit

$$N(x, t) = \sum_{i=0}^{n} \sum_{j=0}^{m} (q_{i,j} o_{i,j}^5). \tag{5.50}$$

In the above relations the following assumptions are taken into consideration

$$f_i^1 = x^i, \quad f_i^2 = (a - x)^{n-i}, \quad \lambda_i = \frac{1}{a^n} \binom{n}{i}, \quad i = 0, \dots, n$$

$$g_j^1 = t^j, \quad g_j^2 = (b - t)^{m-j}, \quad \gamma_j = \frac{1}{b^m} \binom{m}{j}, \quad j = 0, \dots, m. \tag{5.51}$$

5.4 Simulations

In this section, several real applications are used to show the use of neural networks and Bernstein neural networks to approximate the solutions of FDEs.

Example 5.1 Consider the following Burgers–Fisher problem

$$\frac{\partial u(x, t)}{\partial t} + \frac{1}{100} u(x, t) \frac{\partial u(x, t)}{\partial x} - \frac{\partial^2 u(x, t)}{\partial x^2}$$
$$= \frac{1}{100} u(x, t)(1 - u(x, t)) \tag{5.52}$$

on the domain $(x, t) \in [0, 1] \times [0, 1]$, with initial condition

$$u(x, 0) = \frac{1}{2} + \frac{1}{2} \tanh\left(\frac{-1}{400} x\right) \tag{5.53}$$

Table 5.1 E_1 error analysis of the number of iterations.

m	RBF method	m	NN method $r = 30$
7	$7.6647e - 4$	7	$9.9107e - 4$
11	$4.6132e - 4$	11	$7.0018e - 4$
17	$3.3978e - 4$	17	$6.3419e - 4$
m	NN method $r = 60$	m	NN method $r = 90$
7	$6.0121e - 4$	7	$9.9107e - 5$
11	$8.6132e - 5$	11	$5.5018e - 5$
17	$6.3978e - 5$	17	$3.3419e - 5$
m	NN method $r = 120$	m	NN method $r = 150$
7	$6.6647e - 5$	7	$4.9107e - 5$
11	$3.0132e - 5$	11	$1.0018e - 5$
17	$9.3978e - 6$	17	$7.3419e - 6$

and boundary conditions

$$u(0, t) = \frac{1}{2} + \frac{1}{2} \tanh\left(\frac{1}{400}\left(\frac{401}{200}t\right)\right)$$
$$u(1, t) = \frac{1}{2} + \frac{1}{2} \tanh\left(\frac{1}{400}\left(\frac{401}{200}t - 1\right)\right). \tag{5.54}$$

The exact solution for the given problem can be stated as

$$u(x, t) = \frac{1}{2} + \frac{1}{2} \tanh\left(\frac{1}{400}\left(\frac{401}{200}t - x\right)\right). \tag{5.55}$$

For the training of the network, a back propagation learning algorithm is implemented for discretization parameters $h = h'$ and momentum constant $\gamma = 0.1$ on various training steps. In Table 5.1, some values of the least square errors $(E_1 = \sqrt{\frac{1}{n}\sum_{i=1}^{n} |u(x_i, t_i) - u_m(x_i, t_i)|})$ are given. The proposed algorithm on different training steps and the radial basis functions (RBFs) method [187] are compared considering the shape parameter as $c = 1.5$. ∎

Example 5.2 Consider the following Buckley–Leverett differential equation problem

$$\frac{\partial u(x, t)}{\partial t} + \frac{\partial f(u(x, t))}{\partial x} = \frac{\partial^2 A(u(x, t))}{\partial x^2} \tag{5.56}$$

where

$$f(u) = \frac{u^2}{u^2 + (1-u)^2}, \quad a(u) = 4\varepsilon u(1-u) \tag{5.57}$$

on the domain $(x, t) \in [0, 1] \times (0, 0.5)$, with initial condition

$$u_0(x) = \begin{cases} 0 \\ 1 \end{cases} \tag{5.58}$$

boundary conditions

$$u(0, t) = 1, \quad u(1, t) = 0 \tag{5.59}$$

and $\varepsilon = 0.01$. The following specifications are considered.

i) Time step: $h' = 0.98 \dfrac{h^2}{h\|f'\|_\infty + 2\varepsilon\|a\|_\infty}$.

ii) L^1 error: $E_{\text{mid}} = \left(\sum\right)_{i=1}^{n} \sum_{j=1}^{n'} |u(x_i, t_j)|^{-1} \sum_{i=1}^{n} \sum_{j=1}^{n'} |u(x_i, t_j) - u_m(x_i, t_j)|$, where $u_m(x_i, t_j)$ and $u(x_i, t_j)$ are the computed solution and exact value of the reference solution at grid point (x_i, t_j), respectively.

This problem is solved by utilizing the technique of neural networks. The obtained numerical results are displayed in Table 5.2. Comparisons between the proposed algorithm on different training steps and the discrete mollification method with support parameter $\gamma = (m-1)/2$ [12 80] are illustrated. It is clear that the neural network method is effective when compared with the discrete mollification method. ■

Example 5.3   Consider the following strongly degenerate parabolic equation

$$\frac{\partial u(x,t)}{\partial t} + \frac{\partial f(u(x,t))}{\partial x} = \frac{\partial^2 A(u(x,t))}{\partial x^2}, \quad (x,t) \in [0,1] \times [0,1] \tag{5.60}$$

with

$$f(u) = u - u^2, \quad a(u) = \begin{cases} 0 \\ 1 \end{cases}, \quad u_0(x) = \begin{cases} 1 \\ 0 \end{cases}. \tag{5.61}$$

The initial network parameters are chosen based on assumptions made in the previous example. The iterative process gives the results, which are compared with the results obtained from the discrete mollification scheme [12, 80]. Table 5.3 gives approximate errors obtained from the neural network method and the discrete mollification method. Increasing the number of training steps gives higher efficiency. ■

Example 5.4   The following wave equation models the motion of a guitar string of length L

$$\frac{\partial^2 v(x,t)}{\partial t^2} = c^2 \frac{\partial^2 v(x,t)}{\partial x^2} \tag{5.62}$$

Table 5.2 L^1 error analysis of the number of iterations.

	Mollified scheme			NN method		
				r = 30		
1/h	γ = 3	γ = 5	γ = 8	m = 7	m = 11	m = 17
64	$2.6105e - 2$,	$2.5327e - 2$,	$2.5055e - 2$	$2.9107e - 2$,	$2.7166e - 2$,	$2.6921e - 2$
128	$1.4932e - 2$,	$1.4287e - 2$,	$1.4133e - 2$	$2.0018e - 2$,	$1.8203e - 2$,	$1.6104e - 2$
256	$8.3709e - 3$,	$7.9698e - 3$,	$7.6883e - 3$	$1.3419e - 2$,	$1.2051e - 2$,	$1.0918e - 2$
512	$4.5075e - 3$,	$4.3271e - 3$,	$4.1141e - 3$	$9.2177e - 3$,	$9.0027e - 3$,	$8.9723e - 3$
1024	$1.9997e - 3$,	$1.9335e - 3$,	$1.8279e - 3$	$8.0180e - 3$,	$7.9217e - 3$,	$7.9012e - 3$

	NN method			NN method		
		r = 60			r = 90	
1/h	m = 7	m = 11	m = 17	m = 7	m = 11	m = 17
64	$7.2745e - 3$,	$6.5306e - 3$,	$5.2570e - 3$	$9.3490e - 4$,	$8.1529e - 4$,	$7.1160e - 4$
128	$5.3106e - 3$,	$5.6850e - 3$,	$3.9183e - 3$	$7.4572e - 4$,	$6.0627e - 4$,	$5.3375e - 4$
256	$3.4590e - 3$,	$3.1800e - 3$,	$2.6377e - 3$	$5.2688e - 4$,	$4.1019e - 4$,	$3.3007e - 4$
512	$2.2185e - 3$,	$2.0087e - 3$,	$1.7991e - 3$	$4.8935e - 4$,	$3.0067e - 4$,	$2.3441e - 4$
1024	$1.1138e - 3$,	$8.5540e - 4$,	$7.0069e - 4$	$3.2527e - 4$,	$2.8699e - 4$,	$1.0562e - 4$

	NN method			NN method		
		r = 120			r = 150	
1/h	m = 7	m = 11	m = 17	m = 7	m = 11	m = 17
64	$2.0076e - 4$,	$1.2190e - 4$,	$9.2097e - 5$	$5.8729e - 5$,	$4.0680e - 5$,	$3.6279e - 5$
128	$9.6320e - 5$,	$8.65049e - 5$,	$7.9782e - 5$	$4.8796e - 5$,	$3.3274e - 5$,	$2.1607e - 5$
256	$8.1674e - 5$,	$7.1264e - 5$,	$6.1739e - 5$	$3.0508e - 5$,	$2.3466e - 5$,	$1.2830e - 5$
512	$7.9793e - 5$,	$6.4361e - 5$,	$5.0027e - 5$	$2.7411e - 5$,	$1.0087e - 5$,	$8.8939e - 6$
1024	$6.3581e - 5$,	$5.0207e - 5$,	$4.0511e - 5$	$1.3481e - 5$,	$8.2733e - 6$,	$6.2283e - 6$

with the boundary conditions on the domain $(x, t) \in [0, L] \times [0, T]$

$$v(0, t) = \sin(\pi t), \quad v(L, t) = 0 \qquad (5.63)$$

and initial position and velocity

$$v(x, 0) = 0, \quad v_t(x, 0) = \pi \cos(\pi x). \qquad (5.64)$$

Table 5.3 L^1 error analysis of the number of iterations of a strongly degenerate parabolic equation.

	Mollified scheme			ANN method $r = 30$		
$1/h$	$\gamma = 3$	$\gamma = 5$	$\gamma = 8$	$m = 7$	$m = 11$	$m = 17$
16	$6.1071e-2,$	$6.1339e-2,$	$6.2973e-2$	$6.4118e-2,$	$6.4522e-2,$	$6.5601e-2$
32	$2.9857e-2,$	$2.9991e-2,$	$3.0526e-2$	$3.2188e-2,$	$3.3366e-2,$	$3.4928e-2$
64	$1.5843e-2,$	$1.6003e-2,$	$1.6154e-2$	$1.9021e-2,$	$1.9866e-2,$	$2.0021e-2$
128	$7.2514e-3,$	$7.3764e-3,$	$7.5281e-3$	$1.5144e-2,$	$1.6264e-2,$	$1.7651e-2$
256	$3.3466e-3,$	$3.4070e-3,$	$3.4808e-3$	$4.1181e-3,$	$4.2121e-3,$	$4.3652e-3$
512	$1.3855e-3,$	$1.4149e-3,$	$1.4512e-3$	$2.9154e-3,$	$2.9554e-3,$	$2.9851e-3$

	ANN method $r = 60$			ANN method $r = 90$		
$1/h$	$m = 7$	$m = 11$	$m = 17$	$m = 7$	$m = 11$	$m = 17$
16	$2.0254e-3,$	$2.1199e-3,$	$2.2574e-3$	$7.3490e-5,$	$9.0224e-5,$	$1.1880e-4$
32	$1.1122e-3,$	$1.2191e-3,$	$1.2875e-3$	$6.2851e-5,$	$8.7774e-5,$	$9.9889e-5$
64	$8.5154e-4,$	$9.0154e-4,$	$9.6541e-4$	$4.7798e-5,$	$6.8872e-5,$	$7.5541e-5$
128	$5.6213e-4,$	$6.2301e-4,$	$7.0112e-4$	$3.2241e-5,$	$5.9121e-5,$	$6.1004e-5$
256	$3.1656e-4,$	$4.1533e-4,$	$5.0054e-4$	$2.5014e-5,$	$4.6102e-5,$	$5.3236e-5$
512	$8.0228e-5,$	$1.5254e-4,$	$2.4054e-4$	$1.0054e-5,$	$3.1112e-5,$	$4.6225e-5$

	ANN method $r = 120$			ANN method $r = 150$		
$1/h$	$m = 7$	$m = 11$	$m = 17$	$m = 7$	$m = 11$	$m = 17$
16	$9.8874e-6,$	$1.3371e-5,$	$2.2019e-5$	$2.7438e-6,$	$2.9699e-6,$	$4.5308e-6$
32	$8.1259e-6,$	$8.8841e-6,$	$1.0009e-5$	$1.0041e-6,$	$1.3659e-6,$	$2.7257e-6$
64	$7.2342e-6,$	$7.8995e-6,$	$9.0234e-6$	$8.8713e-7,$	$9.2155e-7,$	$1.0009e-6$
128	$6.2248e-6,$	$6.5131e-6,$	$8.1001e-6$	$6.531e-7,$	$7.1365e-7,$	$8.8872e-7$
256	$5.0021e-6,$	$5.3232e-6,$	$7.3634e-6$	$5.3119e-7,$	$6.2241e-7,$	$7.1655e-7$
512	$4.4481e-6,$	$4.7661e-6,$	$6.1301e-6$	$4.1037e-7,$	$5.1093e-7,$	$6.0621e-7$

In the proposed problem $c^2 = \frac{T_s}{\rho}$, T_s is the tension in the string, and ρ is the density of the string. The specifications are $L = 1$, $T = 4$, $T_s = 2\frac{N}{m}$, and $\rho = 2\frac{kg}{m^3}$. The exact solution of the problem is $v(x,t) = \cos(\pi x)\sin(\pi t)$.

The static Bernstein neural network and the dynamic Bernstein neural network are used to approximate the solution. To compare the results, two other

Table 5.4 Approximation errors of different methods.

x	Three-point explicit	Optimal explicit	SNN	DNN
0.1	$1.5e-3$	$3.3e-5$	$1.2e-5$	$2.4e-6$
0.2	$1.4e-3$	$3.0e-5$	$1.4e-5$	$2.3e-6$
0.3	$1.7e-3$	$3.2e-5$	$1.1e-5$	$2.5e-6$
0.4	$1.6e-3$	$3.1e-5$	$1.6e-5$	$2.2e-6$
0.5	$1.5e-3$	$3.3e-5$	$1.7e-5$	$2.7e-6$
0.6	$1.5e-3$	$3.4e-5$	$1.5e-5$	$2.6e-6$
0.7	$1.9e-3$	$3.1e-5$	$1.3e-5$	$2.4e-6$
0.8	$1.8e-3$	$3.2e-5$	$1.4e-5$	$2.8e-6$
0.9	$1.7e-3$	$3.4e-5$	$1.2e-5$	$2.2e-6$
1.0	$1.6e-3$	$3.2e-5$	$1.6e-5$	$2.5e-6$

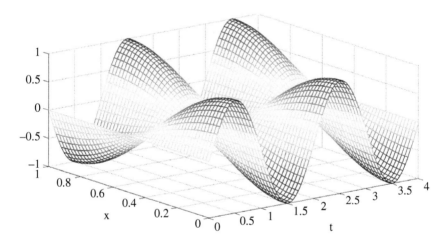

Figure 5.2 Plot of the exact solution.

popular methods are used: the three-point explicit method and the optimal explicit method [63]. The results are compared in Table 5.4. The exact solution is illustrated in Figure 5.2. Corresponding approximated error plots are shown in Figure 5.3. ∎

Example 5.5 Two semi-infinite strings of different densities are joined as [188]

$$\frac{\partial^2 v(x,t)}{\partial t^2} = (c_1^2 + c_2^2)\frac{\partial^2 v(x,t)}{\partial x^2} \tag{5.65}$$

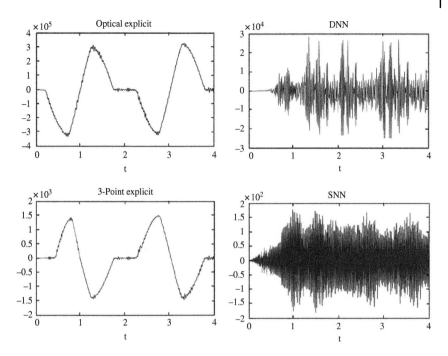

Figure 5.3 Plot of the approximated error using the three-point explicit, optimal explicit, SNN, and DNN methods.

Figure 5.4 Two semi-infinite strings of different densities.

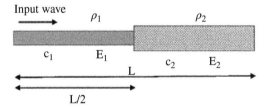

with the boundary conditions on the domain $(x, t) \in [0, L] \times [0, T]$

$$v(0, t) = cos(\pi t), \quad v(L, t) = 0 \qquad (5.66)$$

and the initial position and velocity are

$$v(x, 0) = cos(\pi x), \quad v_t(x, 0) = 0 \qquad (5.67)$$

see Figure 5.4. In the proposed problem $c = \sqrt{\frac{E}{\rho}}$, E is Young's modulus and ρ is the density of the rod. The specifications are $L = 1$, $T = 5$, $E_1 = 2$ kg ms^{-2}, $\rho_1 = 2.882$ kg m^{-3}, $E_2 = 4.3$ kg ms^{-2} and $\rho_1 = 15.136$ kg m^{-3}. The exact solution

Table 5.5 Approximation errors of the Bernstein neural networks.

r	SNN method	DNN method
10	$6.2458e-2$	$3.3193e-2$
20	$1.2572e-2$	$6.5301e-3$
30	$5.9854e-3$	$8.3357e-4$
40	$9.0584e-4$	$1.8527e-4$
50	$5.5487e-4$	$6.0024e-5$
60	$8.0125e-5$	$9.1254e-6$
70	$1.2561e-5$	$3.0125e-6$

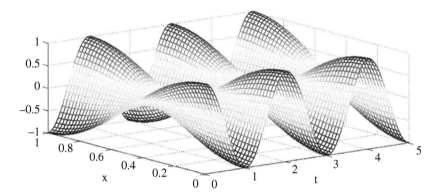

Figure 5.5 Plot of the exact solution.

of the problem is illustrated as

$$v(x,t) = \frac{1}{2}(\cos(\pi(x+t)) + \cos(\pi(x-t))). \tag{5.68}$$

The errors that are obtained from the static Bernstein neural network and dynamic Bernstein neural network are illustrated in Table 5.5. The exact solution is illustrated in Figure 5.5. Figure 5.6 shows the approximated errors with the static Bernstein neural network and the dynamic Bernstein neural network. ∎

Figure 5.6 Plot of the approximated error using the SNN and DNN for *r* = 70.

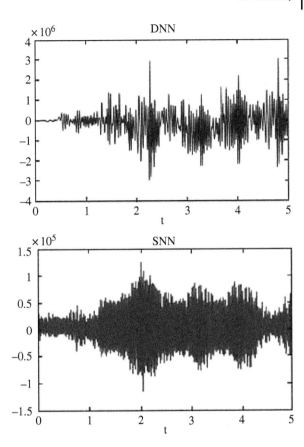

5.5 Summary

In this chapter, solutions to the strongly degenerate parabolic and Burgers–Fisher equations are approximated by neural networks. Furthermore, two types of Bernstein neural network are used for solving a wave equation. The theoretical analysis and simulation results show that these new methods, neural networks and Bernstein neural networks, are effective in estimating solutions to PDEs.

6

System Control using Z-numbers

6.1   Introduction

In this chapter, dual fuzzy equations [192] and fuzzy differential equations (FDEs) are used to model uncertain nonlinear systems, where the coefficients are Z-numbers. The existence of solutions to the dual fuzzy equations and FDEs is discussed. This corresponds to the controllability problem of fuzzy control [58]. Two types of neural networks, feed forward, and feed back networks, are used to approximate the solutions (control actions) to the dual fuzzy equations and FDEs. Several real examples are used to show the effectiveness of the fuzzy control design methods.

6.2   Modeling using Dual Fuzzy Equations and Z-numbers

The decisions are carried out on the basis of knowledge. In order to make the decision significant, the knowledge acquired must be credible. Z-numbers are connected to the reliability of knowledge [198]. Many fields related to the analysis of decisions actually use the idea of Z-numbers. Z-numbers are much less complex to calculate compared with nonlinear system modeling methods. Z-numbers are an abundantly adequate number compared with fuzzy numbers. Although Z-numbers are implemented in much of the literature, from a theoretical point of view this approach has not yet been certified completely.

The Z-number is a new idea that is referred to a higher potential in order to illustrate the information of a human being and to use it in information processing [96][100][101][168][198]. Z-numbers can be considered as being able to answer questions and carry out decisions [110]. There are a few structures based on the theoretical concept of Z-numbers [72]. In [21], an inception that results in the extension of the Z-number is given. In [111], a theorem to transfer Z-numbers to the usual fuzzy sets is proposed.

The following definitions will be used in this chapter.

Modeling and Control of Uncertain Nonlinear Systems with Fuzzy Equations and Z-Number,
First Edition. Wen Yu and Raheleh Jafari.

Definition 6.1 (Z-numbers) *A Z-number has two components* $Z = [A(x), p]$. *The primary component* $A(x)$ *is termed a restriction on a real valued uncertain variable x. The secondary component p is a measure of the reliability of A. p can be reliability, strength of belief, probability, or possibility. When* $A(x)$ *is a fuzzy number and p is the probability distribution of x, the Z-number is defined as a* Z^+-*number. When both* $A(x)$ *and p are fuzzy numbers, the Z-number is defined as a* Z^--*number.*

The Z^+-number has more information than the Z^--number. Here the definition of a Z^+-number is used, i.e. $Z = [A, p]$, A is a fuzzy number, p is a probability distribution.

The membership functions are used to express the fuzzy number. The most popular membership functions are the triangular function

$$\mu_A = F(a, b, c) = \begin{cases} \frac{x-a}{b-a} & a \le x \le b \\ \frac{c-x}{c-b} & b \le x \le c \end{cases} \quad \text{otherwise} \quad \mu_A = 0 \tag{6.1}$$

and trapezoidal function

$$\mu_A = F(a, b, c, d) = \begin{cases} \frac{x-a}{b-a} & a \le x \le b \\ \frac{d-x}{d-c} & c \le x \le d \\ 1 & b \le x \le c \end{cases} \quad \text{otherwise} \quad \mu_A = 0 \tag{6.2}$$

The probability measure is illustrated as

$$P = \int_R \mu_A(x) p(x) dx \tag{6.3}$$

where p is the probability density of x and R is the restriction on p. For discrete Z-numbers

$$P(A) = \sum_{i=1}^{n} \mu_A(x_i) p(x_i). \tag{6.4}$$

Definition 6.2 (α-level of Z-numbers) *The α-level of the Z-number* $Z = (A, P)$ *is demonstrated as*

$$[Z]^\alpha = ([A]^\alpha, [p]^\alpha) \tag{6.5}$$

where $0 < \alpha \le 1$. $[p]^\alpha$ *is calculated by the Nguyen theorem*

$$[p]^\alpha = p([A]^\alpha) = p\left(\left[\underline{A}^\alpha, \overline{A}^\alpha\right]\right) = \left[\underline{P}^\alpha, \overline{P}^\alpha\right] \tag{6.6}$$

where $p([A]^\alpha) = \{p(x) | x \in [A]^\alpha\}$. *So* $[Z]^\alpha$ *can be illustrated as the α-level of a fuzzy number*

$$[Z]^\alpha = \left(\underline{Z}^\alpha, \overline{Z}^\alpha\right) = \left(\left(\underline{A}^\alpha, \underline{P}^\alpha\right), \left(\overline{A}^\alpha, \overline{P}^\alpha\right)\right) \tag{6.7}$$

where $\underline{P}^\alpha = \underline{A}^\alpha p(\underline{x}_i^\alpha)$, $\overline{P}^\alpha = \overline{A}^\alpha p(\overline{x}_i^\alpha)$, *and* $[x_i]^\alpha = (\underline{x}_i^\alpha, \overline{x}_i^\alpha)$.

Definition 6.3 (Supremum metrics for Z-numbers) [22] *The supremum metrics in $\hat{Z}^n$ are suggested as*

$$D(Z_1, Z_2) = d(A_1, A_2) + d(P_1, P_2). \qquad (6.8)$$

In this case $d(\cdot, \cdot)$ are the supremum metrics of fuzzy sets [117]. $(\hat{Z}^n, D)$ is a complete metric space. $D(Z_1, Z_2)$ has the following properties

$$
\begin{aligned}
D(Z_1 + Z, Z_2 + Z) &= D(Z_1, Z_2) \\
D(Z_2, Z_1) &= D(Z_1, Z_2) \\
D(\lambda Z_1, \lambda Z_2) &= |\lambda| D(Z_1, Z_2), \quad \lambda \in R \\
D(Z_1, Z_2) &\leq D(Z_1, Z) + D(Z, Z_2).
\end{aligned}
\qquad (6.9)
$$

Similar to fuzzy numbers [95], Z-numbers also have four primary operations: $\oplus, \ominus, \odot$, and $\oslash$. These operations are demonstrated by sum, subtract, multiply, and division. The operations in this chapter are different definitions with [199]. The α-level of Z-numbers is applied to simplify the operations.

Let us consider $Z_1 = (A_1, p_1)$ and $Z_2 = (A_2, p_2)$ to be two discrete Z-numbers illustrating the uncertain variables x_1 and x_2, $\sum_{k=1}^{n} p_1(x_{1k}) = 1$, and $\sum_{k=1}^{n} p_2(x_{2k}) = 1$. The operations are defined as

$$Z_{12} = Z_1 * Z_2 = (A_1 * A_2, p_1 * p_2) \qquad (6.10)$$

where $* \in \{\oplus, \ominus, \odot, \oslash\}$.

The operations for the fuzzy numbers are defined as [95]

$$
\begin{aligned}
[A_1 \oplus A_2]^{\alpha} &= \left(\underline{A_1}^{\alpha} + \underline{A_2}^{\alpha}, \overline{A_1}^{\alpha} + \overline{A_2}^{\alpha} \right) \\
[A_1 \ominus A_2]^{\alpha} &= \left[\underline{A_1}^{\alpha} - \underline{A_2}^{\alpha}, \overline{A_1}^{\alpha} - \overline{A_2}^{\alpha} \right] \\
[A_1 \odot A_2]^{\alpha} &= \left(\underline{A_1}^{\alpha}\underline{A_2}^{\alpha} + \underline{A_1}^{\alpha}\underline{A_2}^{\alpha} - \underline{A_1}^{\alpha}\underline{A_2}^{\alpha}, \overline{A_1}^{\alpha}\overline{A_2}^{\alpha} + \overline{A_1}^{\alpha}\overline{A_2}^{\alpha} - \overline{A_1}^{\alpha}\overline{A_2}^{\alpha} \right).
\end{aligned}
\qquad (6.11)
$$

For all $p_1 * p_2$ operations, the convolution for the discrete probability distributions is used

$$p_1 * p_2 = \sum_i p_1(x_{1,i}) p_2(x_{2,(n-i)}) = p_{12}(x). \qquad (6.12)$$

If A is a triangle function, the absolute value of the Z-number $Z = (A, p)$ is

$$|Z(x)| = (|a_1| + |b_1| + |c_1|, p(|a_2| + |b_2| + |c_2|)). \qquad (6.13)$$

Also the above definitions satisfy the generalized Hukuhara difference [41]

$$Z_1 \ominus_{gH} Z_2 = Z_{12} \iff \begin{cases} (1) & Z_1 = Z_2 \oplus Z_{12} \\ (2) & Z_2 = Z_1 \oplus (-1)Z_{12} \end{cases}. \qquad (6.14)$$

It is convenient to display that (1) and (2) in combination are genuine if and only if Z_{12} is a crisp number. With respect to the α-level the extracted result is $[Z_1 \ominus_{gH} Z_2]^\alpha = [\min\{\underline{Z}_1^\alpha - \underline{Z}_2^\alpha, \overline{Z}_1^\alpha - \overline{Z}_2^\alpha\}, \max\{\underline{Z}_1^\alpha - \underline{Z}_2^\alpha, \overline{Z}_1^\alpha - \overline{Z}_2^\alpha\}]$ and if $Z_1 \ominus_{gH} Z_2$ and $Z_1 \ominus_H Z_2$ subsist, $Z_1 \ominus_H Z_2 = Z_1 \ominus_{gH} Z_2$. The circumstances for the inerrancy of $Z_{12} = Z_1 \ominus_{gH} Z_2 \in E$ are

$$
\begin{cases}
(1) & \underline{Z}_{12}^\alpha = \underline{Z}_1^\alpha - \underline{Z}_2^\alpha \quad \text{and} \quad \overline{Z}_{12}^\alpha = \overline{Z}_1^\alpha - \overline{Z}_2^\alpha \\
& \text{with } \underline{Z}_{12}^\alpha \text{ increasing}, \quad \overline{Z}_{12}^\alpha \text{ decreasing}, \underline{Z}_{12}^\alpha \leq \overline{Z}_{12}^\alpha \\
(2) & \underline{Z}_{12}^\alpha = \overline{Z}_1^\alpha - \overline{Z}_2^\alpha \quad \text{and} \quad \overline{Z}_{12}^\alpha = \underline{Z}_1^\alpha - \underline{Z}_2^\alpha \\
& \text{with } \underline{Z}_{12}^\alpha \text{ increasing}, \quad \overline{Z}_{12}^\alpha \text{ decreasing}, \underline{Z}_{12}^\alpha \leq \overline{Z}_{12}^\alpha
\end{cases}
\tag{6.15}
$$

where $\forall \alpha \in [0, 1]$

Definition 6.4 (α-level of a Z-number valued function) *Let $\tilde{Z}$ denote the space of Z-numbers. The α-level of Z-number valued function $F : [0, a] \to \tilde{Z}$ is*

$$
F(x, \alpha) = [\underline{F}(x, \alpha), \overline{F}(x, \alpha)]
\tag{6.16}
$$

where $x \in \tilde{Z}$, for each $\alpha \in [0, 1]$.

With the definition of the generalized Hukuhara difference, the generalized Hukuhara derivative (gH-derivative) of F at x_0 is illustrated as

$$
\frac{d}{dt} F(x_0) = \lim_{h \to 0} \frac{1}{h} [F(x_0 + h) \ominus_{gH} F(x_0)].
\tag{6.17}
$$

In (6.17), $F(x_0 + h)$ and $F(x_0)$ display a similar style to Z_1 and Z_2, respectively, which is shown in (6.14).

If the α-level (6.5) is applied to $f(t, x)$ in (6.65), then two Z-number valued functions are obtained: $\underline{f}\left[t, \underline{x}(\zeta, \alpha), \overline{x}(\zeta, \alpha)\right]$ and $\overline{f}\left[t, \underline{x}(\zeta, \alpha), \overline{x}(\zeta, \alpha)\right]$.

Now fuzzy equation (3.5) or (3.6) is used to model the uncertain nonlinear system (3.2). The parameters of the fuzzy equation (3.6) are in the form of Z-numbers,

$$
y_k = a_1 \odot f_1(x_k) \oplus a_2 \odot f_2(x_k) \oplus \cdots \oplus a_n \odot f_n(x_k)
\tag{6.18}
$$

or

$$
\begin{aligned}
& a_1 \odot f_1(x_k) \oplus a_2 \odot f_2(x_k) \oplus \cdots \oplus a_n \odot f_n(x_k) \\
& = b_1 \odot g_1(x_k) \oplus b_2 \odot g_2(x_k) \oplus \cdots \oplus b_m \odot g_m(x_k) \oplus y_k
\end{aligned}
\tag{6.19}
$$

where a_i and b_i are Z-numbers.

Taking into consideration a particular case, $f_i(x_k)$ has polynomial pattern,

$$
(a_1 \odot x_k) \oplus \cdots \oplus (a_n \odot x_k^n) = (b_1 \odot x_k) \oplus \cdots \oplus (b_n \odot x_k^n) \oplus y_k
\tag{6.20}
$$

and (3.29) is termed as dual polynomial based on a Z-number.

The objective of the *modeling* is to minimize error between the two outputs y_k and z_k. As y_k is denoted as a Z-number and z_k is considered to be crisp

Z-number, hence the minimum of every point is applied as the model mentioned below

$$\min_k |y_k - z_k| = \min_k |\beta_k|$$
$$y_k = \left((u_1(k), u_2(k), u_3(k)), p\left(v_1(k), v_2(k), v_3(k)\right) \right)$$
$$\beta_k = \left((\rho_1(k), \rho_2(k), \rho_3(k)), p\left(\varphi_1(k), \varphi_2(k), \varphi_3(k)\right) \right). \quad (6.21)$$

By the definition of absolute value (6.13),

$$\min_k |\beta_k| = \min_k [(|u_1(k) - f(x_k)| + |u_2(k) - f(x_k)| + |u_3(k) - f(x_k)|),$$
$$(|p(v_1(k)) - f(x_k)| + |p(v_2(k)) - f(x_k)| + |p(v_3(k)) - f(x_k)|)]$$
$$\rho_1(k) = \min_k |u_1(k) - f(x_k)|, \quad \rho_2(k) = \min_k |u_2(k) - f(x_k)|,$$
$$\rho_3(k) = \min_k |u_3(k) - f(x_k)|$$
$$p(\varphi_1(k)) = \min_k |p(v_1(k)) - f(x_k)|, \quad p(\varphi_2(k)) = \min_k |p(v_2(k)) - f(x_k)|,$$
$$p(\varphi_3(k)) = \min_k |p(v_3(k)) - f(x_k)|. \quad (6.22)$$

The modeling problem (6.21) is to find $u_1(k), u_2(k), u_3(k), p(v_1(k)), p(v_2(k))$, and $p(v_3(k))$ in such a manner that

$$\min_{u_1(k),u_2(k),u_3(k),p(v_1(k)),p(v_2(k)),p(v_3(k))} \left\{ \max_k |\beta_k| \right\}$$
$$= \min_{u_1(k),u_2(k),u_3(k),p(v_1(k)),p(v_2(k)),p(v_3(k))} \left\{ \max_k |y_k - f(x_k)| \right\}. \quad (6.23)$$

Considering (6.22)

$$\rho_1(k) \geq |u_1(k) - f(x_k)|, \quad \rho_2(k) \geq |u_2(k) - f(x_k)|, \quad \rho_3(k) \geq |u_3(k) - f(x_k)|$$
$$p(\varphi_1(k)) \geq |p(v_1(k)) - f(x_k)|, \quad p(\varphi_2(k)) \geq |p(v_2(k)) - f(x_k)|$$
$$p(\varphi_3(k)) \geq |p(v_3(k)) - f(x_k)|. \quad (6.24)$$

Equation (6.23) can be solved by the application of the linear programming methodology,

$$\begin{cases} & \min \rho_1(k) \\ \text{subject}: & \rho_1(k) + \left\{ \left(\sum_{j=0}^n a_j \odot x_k^j \right) \ominus_{gH} \left(\sum_{j=0}^n b_j \odot x_k^j \right) \right\} \geq f(x_k) \\ & \rho_1(k) - \left\{ \left(\sum_{j=0}^n a_j \odot x_k^j \right) \ominus_{gH} \left(\sum_{j=0}^n b_j \odot x_k^j \right) \right\} \geq -f(x_k) \\ & \min \varphi_1(k) \\ \text{subject}: & p(\varphi_1(k)) + \left\{ \left(\sum_{j=0}^n a_j \odot x_k^j \right) \ominus_{gH} \left(\sum_{j=0}^n b_j \odot x_k^j \right) \right\} \geq f(x_k) \\ & p(\varphi_1(k)) - \left\{ \left(\sum_{j=0}^n a_j \odot x_k^j \right) \ominus_{gH} \left(\sum_{j=0}^n b_j \odot x_k^j \right) \right\} \geq -f(x_k) \end{cases} \quad (6.25)$$

$$
\left\{
\begin{array}{l}
\min \rho_2(k) \\
\text{subject}: \ \rho_2(k) - \left[\sum_{j=0}^{n} \underline{a}_j \underline{x}_k^j - \sum_{j=0}^{n} \underline{b}_j \underline{x}_k^j \right] \geq f(x_k) \\
\rho_2(k) \geq 0 \\
\min \varphi_2(k) \\
\text{subject}: \ p(\varphi_2(k)) - \left[\sum_{j=0}^{n} \underline{a}_j \underline{x}_k^j - \sum_{j=0}^{n} \underline{b}_j \underline{x}_k^j \right] \geq f(x_k) \\
p(\varphi_2(k)) \geq 0
\end{array}
\right. \tag{6.26}
$$

$$
\left\{
\begin{array}{l}
\min \rho_3(k) \\
\text{subject}: \ \rho_3(k) - \left[\sum_{j=0}^{n} \overline{a}_j \overline{x}_k^j - \sum_{j=0}^{n} \overline{b}_j \overline{x}_k^j \right] \geq f(x_k) \\
\rho_3(k) \geq 0 \\
\min \varphi_3(k) \\
\text{subject}: \ p(\varphi_3(k)) - \left[\sum_{j=0}^{n} \overline{a}_j \overline{x}_k^j - \sum_{j=0}^{n} \overline{b}_j \overline{x}_k^j \right] \geq f(x_k) \\
p(\varphi_3(k)) \geq 0
\end{array}
\right. \tag{6.27}
$$

where $\underline{a}_j$, $\underline{b}_j$, $\underline{x}_k$, $\overline{a}_j$, $\overline{b}_j$, and $\overline{x}_k$ are explained as mentioned in (6.5). Therefore, a better way of approximating $f(x_k)$ at the point x_k is y_k. The time for approximating the error β_k is minimized.

The object of the *controller* design is to obtain u_k such that the output of the plant y_k can follow a desired output y_k^*,

$$
\min_{u_k} \|y_k - y_k^*\|. \tag{6.28}
$$

The control object can be considered as: detect a solution u_k for the mentioned dual equation on the basis of the Z-number

$$
(a_1 \odot f_1(x_k)) \oplus (a_2 \odot f_2(x_k)) \oplus \cdots \oplus (a_n \odot f_n(x_k))
$$
$$
= (b_1 \odot g_1(x_k)) \oplus (b_2 \odot g_2(x_k)) \oplus \cdots \oplus (b_m \odot g_m(x_k)) \oplus y_k^* \tag{6.29}
$$

where $x_k = [y_{k-1}^T, y_{k-2}^T, \cdots u_k^T, u_{k-1}^T, \cdots]^T$.

6.3 Controllability using Dual Fuzzy Equations

As the main objective of control is to find u_k of (6.19) based on a Z-number, the controllability signifies that the dual fuzzy equation (6.19) has a solution.

The following lemmas are required.

Lemma 6.1 *If the coefficients of the dual equation* (6.19) *are Z-numbers, then the solution u_k satisfies*

$$\left\{ \cap_{j=1}^{n} \text{domain} \left[f_j(x) \right] \right\} \cap \left\{ \cap_{j=1}^{m} \text{domain} \left[g_j(x) \right] \right\} \neq \phi. \tag{6.30}$$

Proof: Assume $u_0 \in \hat{Z}$ is considered to be a solution of (6.19); the dual equation, which relies on Z-numbers, changes as follows

$$(a_1 \odot f_1(u_0)) \oplus \cdots \oplus (a_n \odot f_n(u_0)) = (b_1 \odot g_1(u_0)) \oplus \cdots \oplus (b_m \odot g_m(u_0)) \oplus y_k^*. \tag{6.31}$$

As $f_j(u_0)$ and $g_j(u_0)$ exist, $u_0 \in \text{domain} \left[f_j(x) \right]$, $u_0 \in \text{domain} \left[g_j(x) \right]$. Subsequently, it can be inferred that $u_0 \in \cap_{j=1}^{n} \text{domain} \left[f_j(x) \right] = D_1$, and $u_0 \in \cap_{j=1}^{m} \text{domain} \left[g_j(x) \right] = D_2$. Hence, there exists u_0 in such a manner that $u_0 \in D_1 \cap D_2 \neq \phi$. ∎

Let two Z-numbers $m_0, n_0 \in \hat{Z}$, $m_0 < n_0$. The set $K(x) = \{ x \in \hat{Z}, m_0 \leq x \leq n_0 \}$ and an operator $S : K \to K$ are defined as

$$S(m_0) \geq m_0, \quad S(n_0) \leq n_0. \tag{6.32}$$

In this manner S is condensing and continuous, and also bounded as $S(z) < r(z)$, $z \subset K$, and $r(z) > 0$. $r(z)$ can be considered as the evaluation of z.

Lemma 6.2 *If $n_i = S\left(n_{i-1} \right)$ and $m_i = S\left(m_{i-1} \right)$, $i = 1, 2, \ldots$, and the upper and lower bounds of S are defined as $\bar{s}$ and $\underline{s}$, then*

$$\bar{s} = \lim_{i \to +\infty} n_i, \quad \underline{s} = \lim_{i \to +\infty} m_i, \tag{6.33}$$

and

$$m_0 \leq m_1 \leq \ldots \leq m_n \leq \ldots \leq n_n \leq \ldots \leq n_1 \leq n_0. \tag{6.34}$$

Proof: As long as S is increasing, it is obvious from (6.32) that (6.34) exists. In this case, it is verified that $\{m_i\}$ connects to some $\underline{s} \in \hat{Z}$ and $S(\underline{s}) = \underline{s}$. The set $B = \{ m_0, m_1, m_2, \ldots \}$ is enclosed and $B = S(B) \bigcup \{ m_0 \}$, thus, $r(B) = r(S(B))$ where $r(B)$ denotes the quantification of the non-compactness of B. It is clear from S that $r(B) = 0$, i.e. B is a proportionally compact set. Thus, there exists an outflow of $\{ m_{n_k} \} \subset \{ m_n \}$ in such a manner that $m_{n_k} \to \underline{s}$ for any $\underline{s} \in \hat{Z}$ (take into consideration that $\hat{Z}$ is complete). Distinctly, $m_n \leq \underline{s} \leq n_n$ $(n = 1, 2, \ldots)$. As in the case $p > n_k$, according to the definition of the supremum metrics for Z-numbers, it is shown that $D(\underline{s}, m_p) \leq D(\underline{s}, m_{n_k})$. Hence, $m_p \to \underline{s}$ as $p \to \infty$. Considering the limit $n \to \infty$ on both sides of the equality $m_n = S(m_{n-1})$, $\underline{s} = S(\underline{s})$ is found and as a result S is continuous and K is closed.

Similarly, it can be concluded that $\{ n_n \}$ converges to some $\bar{s} \in \hat{Z}$ and $S(\bar{s}) = \bar{s}$. So, it should be confirmed that $\bar{s}$ and $\underline{s}$ are the maximal and minimal fixed point of S in K, respectively. Assume $\tilde{s} \in K$ and $S(\tilde{s}) = \tilde{s}$. As S is increasing, it is clear from $m_0 \leq \tilde{s} \leq n_0$ that $S(m_0) \leq S(\tilde{s}) \leq S(n_0)$, i.e. $m_1 \leq \tilde{s} \leq n_1$. Utilizing similar logic, it is obtained that $m_2 \leq \tilde{s} \leq n_2$, and formally, $m_n \leq \tilde{s} \leq n_n$ $(n = 1, 2, 3, \ldots)$. Here, considering the limit $n \to \infty$, $\underline{s} \leq \tilde{s} \leq \bar{s}$ is extracted.

The fixed point will result in x_0 inside K, the consecutive iterates $x_i = S\left(x_{i-1}\right)$, $i = 1, 2, \ldots$ will result in convergency towards x_0, i.e. the supremum matrix (6.8) $\lim_{i \to \infty} D(x_i, x_0) = 0$. ∎

Theorem 6.1 *If a_i and b_j $(i = 1, \ldots, n, j = 1, \ldots, m)$ in (6.19) are Z-numbers, and they satisfy the Lipschitz condition*

$$\left| (d_{M_1}(a_i), d_{M_2}(a_i)) - (d_{M_1}(a_k), d_{M_2}(a_k)) \right|$$
$$\leq H \left| a_i(M_1) - a_k(M_1) \right| + H \left| a_i(M_2) - a_k(M_2) \right|$$

$$\left| (d_{U_1}(a_i), d_{U_2}(a_i)) - (d_{U_1}(a_k), d_{U_2}(a_k)) \right|$$
$$\leq H \left| a_i(U_1) - a_k(U_1) \right| + H \left| a_i(U_2) - a_k(U_2) \right| \tag{6.35}$$

and the upper bounds of the functions f_i and g_j are $\left| f_i \right| \leq \bar{f}$, $\left| g_j \right| \leq \bar{g}$, then the dual fuzzy equation (6.19) has a solution u in the following set

$$K_H = \left\{ u \in \tilde{Z}, \left| \overline{u}^{(\alpha_1, \beta_1)} - \underline{u}^{(\alpha_2, \beta_2)} \right| \leq (n\bar{f} \oplus m\bar{g})(H \left| \alpha_1 - \alpha_2 \right| + H \left| \beta_1 - \beta_2 \right|) \right\}. \tag{6.36}$$

Proof: Since a_i and b_j are the Z-numbers, and from the definition (6.35),

$$d_M(\alpha, \beta) = ((a_{1M_1}(\alpha), a_{1M_2}(\beta)) \odot f_1(x)) \oplus \cdots \oplus ((a_{nM_1}(\alpha), a_{nM_2}(\beta)) \odot f_n(x))$$
$$\ominus_{gH}((b_{1M_1}(\alpha), b_{1M_2}(\beta)) \odot g_1(x)) \ominus_{gH} \cdots \ominus_{gH}((b_{mM_1}(\alpha), b_{mM_2}(\beta)) \odot g_m(x)) \tag{6.37}$$

so

$$\left| d_M(\alpha, \beta) - d_M(\varphi, \rho) \right|$$
$$= (\left| f_1(x) \right| \odot \mid (a_{1M_1}(\alpha), a_{1M_2}(\beta)) \ominus_{gH}(a_{1M_1}(\varphi), a_{1M_2}(\rho)) \mid) \oplus \cdots$$
$$\oplus (\left| f_n(x) \right| \odot \mid (a_{nM_1}(\alpha), a_{nM_2}(\beta)) \ominus_{gH}(a_{nM_1}(\varphi), a_{nM_2}(\rho)) \mid)$$
$$\oplus (\left| g_1(x) \right| \odot \mid (b_{1M_1}(\alpha), b_{1M_2}(\beta)) \ominus_{gH}(b_{1M_1}(\varphi), b_{1M_2}(\rho)) \mid) \oplus \cdots$$
$$\oplus (\left| g_m(x) \right| \odot \mid (b_{mM_1}(\alpha), b_{mM_2}(\beta)) \ominus_{gH}(b_{mM_1}(\varphi), b_{mM_2}(\rho)) \mid) \tag{6.38}$$

With respect to the Lipschitz condition (6.35), (6.38) is

$$\left| d_M(\alpha, \beta) - d_M(\varphi, \rho) \right| \leq \bar{f}(H \sum_{i=1}^{n} \left| \alpha - \varphi \right| + H \sum_{i=1}^{n} \left| \beta - \rho \right|) \oplus \bar{g}(H \sum_{i=1}^{m} \left| \alpha - \varphi \right|$$

$$+ H \sum_{i=1}^{m} \left| \beta - \rho \right|) = \left(n\bar{f} \oplus m\bar{g} \right)(H \left| \alpha - \varphi \right| + H \left| \beta - \rho \right|). \tag{6.39}$$

In the same manner, the upper limits satisfy

$$\left| d_U(\alpha, \beta) - d_U(\varphi, \rho) \right| \le \left(n\overline{f} \oplus m\overline{g} \right) (H \left| \alpha - \varphi \right| + H \left| \beta - \rho \right|). \tag{6.40}$$

As the lower limit $\left| d_M(\alpha, \beta) - d_M(\varphi, \rho) \right| \ge 0$, with respect to Lemma 6.2 the solution is in K_H and is defined as (6.36). ∎

Lemma 6.3 *Let us consider the data number to be m and the order of the equation to be n in* (6.20), *also*

$$m \ge 2n + 1 \tag{6.41}$$

where $k = 1, \dots, m$, *hence the solutions of* (6.26) *and* (6.27) *are* $\rho_2(k) = p(\varphi_2(k)) = \rho_3(k) = p(\varphi_3(k)) = 0$.

Proof: Since

$$\sum_{j=0}^{n} \underline{a}_j \underline{x}_k^j - \sum_{j=0}^{n} \underline{b}_j \underline{x}_k^j \le -f(x_k), \qquad i = 1, 2, \dots, m. \tag{6.42}$$

let us choose $2n + 1$ points for x_k, and the result is the interpolation of the dual polynomial based on the Z-number

$$b(k) = \sum_{j=0}^{n} \underline{a}_j \underline{x}_k^j - \sum_{j=0}^{n} \underline{b}_j \underline{x}_k^j. \tag{6.43}$$

Let $h = \max_k \{b(k) + f(x_k)\}$ and $h > 0$. As a result the dual polynomial based onthe Z-number (6.20) can be transformed to the other form of a new dual polynomial based on the Z-number $b(k) - h$. This suggested recent dual polynomial based on the Z-number satisfies (6.42). Therefore, the feasible point of (6.26) $\rho_2(k) \ge 0$ and $p(\varphi_2(k)) \ge 0$, it must be zero. A similar result can be obtained for (6.27). ∎

The solution x_k is a Z-number. In the case of $k = 1 \cdots n$, there should be a validated solution for the equation approximation [126]. Since $u_2(k)$, $v_2(k)$, $u_3(k)$, and $v_3(k)$, (6.25) has a solution.

Theorem 6.2 *If the data number is as big as* (6.41), *and the dual polynomial based on the Z-number* (6.20) *satisfies*

$$D\left[h\left(x_{k1}, u_{k1} \right), h\left(x_{k2}, u_{k2} \right) \right] \le lD\left[u_{k1}, u_{k2} \right] \quad 0 < l < 1 \tag{6.44}$$

where $h(\cdot)$ *represents a dual polynomial based on the Z-number,*

$$h\left(x_{k1}, u_{k1} \right) : (a_1 \odot x_{k1}) \oplus \cdots \oplus (a_n \odot x_{k1}^n)$$
$$= (b_1 \odot x_{k1}) \oplus \cdots \oplus (b_n \odot x_{k1}^n) \oplus y_{k1}. \tag{6.45}$$

$D[u, v]$ *is the Hausdorff distance based on the Z-numbers u and v,*

$$D[u, v] = \max \left\{ \begin{array}{l} \sup_{(x_1, y_1) \in u} \inf_{(x_2, y_2) \in v} \left(d\left(x_1, x_2 \right) + d\left(y_1, y_2 \right) \right) \\ \sup_{(x_1, y_1) \in v} \inf_{(x_2, y_2) \in u} \left(d\left(x_1, x_2 \right) + d\left(y_1, y_2 \right) \right) \end{array} \right\}. \qquad (6.46)$$

If $d\left(x, y \right)$ is the supremum metrics considering fuzzy sets, then (6.20) has a distinct solution u.

Proof: The knowledge that is obtained from Lemma 6.2 states that there exist solutions for (6.25)–(6.27), if it includes too much data that satisfy (6.41). Neglecting lack of generality, it can be considered that the solutions for (6.25)–(6.27) are at par with $x_k = 0$, which tends to u_0. Equation (6.44) signifies $h(\cdot)$ in (6.45) is continuous. If $\delta > 0$ is selected in such a manner that $D[y_k, u_0] \leq \delta$, hence

$$D[h(x_k, u_0), u_0] \leq (1 - l)\delta. \qquad (6.47)$$

Considering $h(0, u_0) = u_0$, x is chosen close to 0, $x_k \in [0, c]$, $c > 0$, and

$$C_0 : \rho = \sup_{x_k \in [0,c]} D[y_{k_1}, y_{k_2}]. \qquad (6.48)$$

Assume $\{y_{k_m}\}$ is a succession in C_0, for any $\varepsilon > 0$ the computation can be done for $N_0(\varepsilon)$ in such a manner that $\rho < \varepsilon, m, n \geq N_0$. Hence $y_{k_m} \rightarrow y_k$ for $x_k \in [0, c]$. Therefore,

$$D[y_k, u_0] \leq D[y_k, y_{k_m}] + D[y_{k_m}, u_0] < \varepsilon + \delta \qquad (6.49)$$

for all $x \in [0, c]$, $m \geq N_0(\varepsilon)$. As $\varepsilon > 0$ is randomly small,

$$D[y_k, u_0] \leq \delta \qquad (6.50)$$

for all $x \in [0, c]$. Now it should be validated that y_k is continuous at $x_0 = 0$. It is given for $\delta > 0$, there exists $\delta_1 > 0$ in such a manner that

$$D[y_k, u_0] \leq D[y_k, y_{k_m}] + D[y_{k_m}, u_0] \leq \varepsilon + \delta_1 \qquad (6.51)$$

for every $m \geq N_0(\varepsilon)$, by means of (6.50), while $|x - x_0| < \delta_1$, y_k is continuous at $x_0 = 0$. As a result (6.20) has a distinct solution u_0. ∎

The necessary circumstance in order to establish the controllability (existence of a solution) of the dual equation, which is based on the Z-number, (6.29) is (6.30), and the sufficient condition of the controllability is (6.35). For the majority of membership functions, such as triangular functions and the trapezoidal function, the Lipschitz condition (6.35) is satisfied. In this case, it is considered to be controllable.

6.4 Fuzzy Controller

There is no analytical solution for the dual fuzzy Equation (6.19). In this section, neural networks are utilized to approximate the solution (control). In order to use neural networks, (6.19) is written as

$$(a_1 \odot f_1(x)) \oplus \cdots \oplus (a_n \odot f_n(x)) \ominus_{gH} (b_1 \odot g_1(x)) \ominus_{gH} \cdots \ominus_{gH} (b_m \odot g_m(x)) = y_k^*.$$
(6.52)

Two types of neural network are used, feed forward and feed back neural networks, to approximate the solutions of (6.52), see Figure 3.2 and Figure 3.3. The inputs to the neural network are the Z-numbers a_i and b_j; the output is the Z-number y_k. The weights are $f_i(x)$ and $g_j(x)$.

The main idea is to find appropriate weights of neural networks such that the output of the neural network $\hat{y}_k$ approaches the desired output y_k^*. From the control point of view, it is a requirement to find a controller u_k that is a function of x such that the output of the plant (3.1) y_k (crisp value) approximates the Z-number y_k^*.

The input Z-numbers a_i and b_i are first applied to α-level as (6.5)

$$[a_i]^\alpha = \left(\underline{a}^\alpha, \overline{a}^\alpha\right) \quad i = 1 \cdots n$$
$$[b_j]^\alpha = \left(\underline{b}^\alpha, \overline{b}^\alpha\right) \quad j = 1 \cdots m$$
(6.53)

Then they are multiplied by the Z-number weights $f_i(x)$ and $g_j(x)$

$$[O_f]^\alpha = \begin{pmatrix} \sum_{i \in M_f} \underline{f_i}^\alpha(x)\underline{a_i}^\alpha + \sum_{i \in C_f} \underline{f_i}^\alpha(x)\overline{a_i}^\alpha, \\ \sum_{i \in M_f'} \overline{f_i}^\alpha(x)\overline{a_i}^\alpha, \sum_{i \in C_f'} \overline{f_i}^\alpha(x)\underline{a_i}^\alpha \end{pmatrix}$$
$$[O_g]^\alpha = \begin{pmatrix} \sum_{j \in M_g} \underline{g_j}^\alpha(x)\underline{b_j}^\alpha + \sum_{j \in C_g} \underline{g_j}^\alpha(x)\overline{b_j}^\alpha, \\ \sum_{j \in M_g'} \overline{g_j}^\alpha(x)\overline{b_j}^\alpha, \sum_{j \in C_g'} \overline{g_j}^\alpha(x)\underline{b_j}^\alpha \end{pmatrix}$$
(6.54)

where $M_f = \{i | \underline{f_i}^\alpha(x) \geq 0\}$, $C_f = \{i | \underline{f_i}^\alpha(x) < 0\}$, $M_f' = \{i | \overline{f_i}^\alpha(x) \geq 0\}$, $C_f' = \{i | \overline{f_i}^\alpha(x) < 0\}$, $M_g = \{j | \underline{g_j}^\alpha(x) \geq 0\}$, $C_g = \{j | \underline{g_j}^\alpha(x) < 0\}$, $M_g' = \{j | \overline{g_j}^\alpha(x) \geq 0\}$, $C_g' = \{j | \overline{g_j}^\alpha(x) < 0\}$.

The neural network output is

$$[\hat{y}_k]^\alpha = \left(\underline{O_f}^\alpha - \underline{O_g}^\alpha, \overline{O_f}^\alpha - \overline{O_g}^\alpha\right).$$
(6.55)

The training error is

$$e_k = y_k^* \ominus \hat{y}_k.$$
(6.56)

Here $[y_k^*]^\alpha = \left(\underline{y_k^*}^\alpha, \overline{y_k^*}^\alpha\right)$, $[\hat{y}_k]^\alpha = \left(\underline{\hat{y}_k}^\alpha, \overline{\hat{y}_k}^\alpha\right)$, and $[e_k]^\alpha = \left(\underline{e_k}^\alpha, \overline{e_k}^\alpha\right)$.

In order to train the weights, a cost function based on the Z-numbers is required as

$$J_k = \underline{J}^\alpha + \overline{J}^\alpha, \quad \underline{J}^\alpha = \frac{1}{2}\left(\underline{y_k^*}^\alpha - \underline{\hat{y}_k}^\alpha\right)^2, \quad \overline{J}^\alpha = \frac{1}{2}\left(\overline{y_k^*}^\alpha - \overline{\hat{y}_k}^\alpha\right)^2.$$
(6.57)

It is clear that $J_k \to 0$ when $[\hat{y}_k]^\alpha \to [y_k^*]^\alpha$.

The index (6.57) is least mean square. It has a self-correcting feature that permits it to operate for an arbitrarily long period without deviating from its constraints. The gradient algorithm is susceptible to cumulative round-off errors.

The gradient technique is utilized to train the Z-number weights $f_i(x)$ and $g_j(x)$. The solution x_0 is a function of $f_i(x)$ and $g_j(x)$. $\frac{\partial J_k}{\partial x_0}$ and $\frac{\partial J_k}{\partial \overline{x_0}}$ are computed as

$$\frac{\partial J_k}{\partial x_0} = \frac{\partial \underline{J}^\alpha}{\partial x_0} + \frac{\partial \overline{J}^\alpha}{\partial x_0}$$
$$\frac{\partial J_k}{\partial \overline{x_0}} = \frac{\partial \underline{J}^\alpha}{\partial \overline{x_0}} + \frac{\partial \overline{J}^\alpha}{\partial \overline{x_0}}$$
(6.58)

According to the chain rule

$$\frac{\partial \underline{J}^\alpha}{\partial x_0} = \frac{\partial \underline{J}^\alpha}{\partial \hat{y}_k^\alpha}\frac{\partial \hat{y}_k^\alpha}{\partial O_f^\alpha}\frac{\partial O_f^\alpha}{\partial f_i^\alpha(x)}\frac{\partial f_i^\alpha(x)}{\partial x_0} - \frac{\partial \underline{J}^\alpha}{\partial \hat{y}_k^\alpha}\frac{\partial \hat{y}_k^\alpha}{\partial O_g^\alpha}\frac{\partial O_g^\alpha}{\partial g_j^\alpha(x)}\frac{\partial g_j^\alpha(x)}{\partial x_0}$$
(6.59)

so

$$\frac{\partial \underline{J}^\alpha}{\partial x_0} = \sum_{i=1}^{n} -\left(y_k^{*\alpha} - \hat{y}_k^\alpha\right)\underline{a}_i^\alpha \underline{f}_i'^\alpha + \sum_{j=1}^{m}\left(y_k^{*\alpha} - \hat{y}_k^\alpha\right)\underline{b}_j^\alpha \underline{g}_j'^\alpha$$
(6.60)

or

$$\frac{\partial \overline{J}^\alpha}{\partial x_0} = \sum_{i=1}^{n} -\left(y_k^{*\alpha} - \hat{y}_k^\alpha\right)\overline{a}_i^\alpha \overline{f}_i'^\alpha + \sum_{j=1}^{m}\left(y_k^{*\alpha} - \hat{y}_k^\alpha\right)\overline{b}_j^\alpha \overline{g}_j'^\alpha,$$
(6.61)

$\frac{\partial J_k}{\partial x_0}$ can be calculated in the same way as above.

The solution x_0 is updated as

$$x_0(k+1) = x_0(k) - \eta\frac{\partial J_k}{\partial x_0}$$
$$\overline{x_0}(k+1) = \overline{x_0}(k) - \eta\frac{\partial J_k}{\partial \overline{x_0}}$$
(6.62)

where η is the rate of the training $\eta > 0$.

In order to increase training process, a momentum term is added as

$$x_0(k+1) = x_0(k) - \eta\frac{\partial J_k}{\partial x_0} + \gamma\left[x_0(k) - x_0(k-1)\right]$$
$$\overline{x_0}(k+1) = \overline{x_0}(k) - \eta\frac{\partial J_k}{\partial \overline{x_0}} + \gamma\left[\overline{x_0}(k) - \overline{x_0}(k-1)\right]$$
(6.63)

where $\gamma > 0$. After x_0 is updated, it should be substituted into the weights $f_i(x_0)$ and $g_j(x_0)$.

The solution of the dual equation is on the basis of the Z-number (6.19), which can also be estimated by a feed back neural network, as in Figure 3.3. In this form, the inputs are the Z-number functions $f_i(x)$ and $g_j(x)$, and the weights are taken to be Z-numbers a_i and b_j. The training error e_k is utilized in order to update x. Once the nonlinear operations $f_i(x)$ and $g_j(x)$ are performed, O_f and O_g are similar to (6.54). The output related to the neural network is similar to (6.55).

6.5 Nonlinear System Modeling

Consider the following controlled unknown nonlinear system

$$\dot{x} = f_1(x_1, u, t) \tag{6.64}$$

where $f_1(x_1, u)$ is an unknown vector function, $x_1 \in \mathfrak{R}^n$ is an internal state vector and $u \in \mathfrak{R}^m$ is the input vector.

Here, the following differential equation (FDE) is used to model the uncertain nonlinear system (6.64),

$$\frac{\mathrm{d}}{\mathrm{d}t}x = f(x, u) \tag{6.65}$$

where $x \in \mathfrak{R}^n$ is the Z-number variable that corresponds to the state x_1 in (6.64), $f(t, x)$ is a Z-number vector function that relates to $f_1(x_1, u)$, and $\frac{\mathrm{d}}{\mathrm{d}t}x$ is the derivative based on the Z-number variable. Here the uncertainties of the nonlinear system (6.64) are in the form of Z-numbers.

The FDE (6.65) can be equivalent to the following four ODEs

$$\begin{array}{l}(1) \begin{cases} \frac{\mathrm{d}}{\mathrm{d}t}\underline{x} = \underline{f}\left[t, \underline{x}(\zeta, \alpha), \overline{x}(\zeta, \alpha)\right] \\ \frac{\mathrm{d}}{\mathrm{d}t}\overline{x} = \overline{f}\left[t, \underline{x}(\zeta, \alpha), \overline{x}(\zeta, \alpha)\right] \end{cases} \\ (2) \begin{cases} \frac{\mathrm{d}}{\mathrm{d}t}\underline{x} = \overline{f}\left[t, \underline{x}(\zeta, \alpha), \overline{x}(\zeta, \alpha)\right] \\ \frac{\mathrm{d}}{\mathrm{d}t}\overline{x} = \underline{f}\left[t, \underline{x}(\zeta, \alpha), \overline{x}(\zeta, \alpha)\right] \end{cases} \end{array} . \tag{6.66}$$

Here, the FDE (6.65) is used to model the uncertain nonlinear system (6.64) such that the output of the plant x can follow the plant output x_1,

$$\min_{f} \|x - x_1\| . \tag{6.67}$$

This modeling object can be considered as: finding $\overline{f}$ and $\underline{f}$ in the fuzzy equations of (6.66) or finding the solutions of these models. It is impossible to obtain analytical solutions. Neural networks are used to approximate them, see Figure 6.1.

In fact, the nonlinear system can be modeled by the neural network directly. However, this data driven black box identification method does not use the model information.

6.6 Controllability using Fuzzy Differential Equations

Theorem 6.3 *If a Z-number function f and its derivative $\frac{\partial f}{\partial x}$ are on the rectangle $[-p, p] \times [-q, q]$, where $p, q \in \tilde{Z}$, $\tilde{Z}$ is the space of the Z-number, then there exists a unique Z-number solution for the following FDE based on Z-numbers*

$$\frac{\mathrm{d}}{\mathrm{d}t}x = f(t, x), \quad x(t_0) = x_0 \tag{6.68}$$

for all $t \in (-b, b)$, $b \leq p$

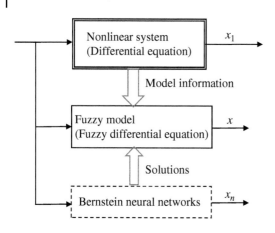

Figure 6.1 Nonlinear system modeling with a fuzzy differential equation.

Proof: Picard's iteration technique [43] is utilized to develop a sequence of Z-number functions $\varphi_n(t)$ as

$$\varphi_{n+1}(t) = \varphi_0 \oplus \int_0^t f(s, \varphi_n(s)) ds$$

$$= \varphi_0 \ominus_H (-1) \int_0^t f(s, \varphi_n(s)) ds. \tag{6.69}$$

Initially, it is validated that $\varphi_n(t)$ is continuous and exist for all n. Obviously, if $\varphi_n(t)$ exists, then $\varphi_{n+1}(t)$ also exists as

$$\varphi_{n+1}(t) = \varphi_0 \oplus \int_0^t f(s, \varphi_n(s)) ds$$

$$= \varphi_0 \ominus_H (-1) \int_0^t f(s, \varphi_n(s)) ds \tag{6.70}$$

Since f is continuous, so there exists $N \in E$ such that $|f(t, x)| \leq N$ for all $t \in [-p, p]$ and all $x \in [-q, q]$. By considering $t \in [-b, b]$ for $b \leq \min(q/N, p)$, then it is possible that

$$\| \varphi_{n+1} \ominus \varphi_0 \| = \left\| \int_0^t f(s, \varphi_n(s)) ds \right\| \leq N|t| \leq Nb \leq q. \tag{6.71}$$

This validates that $\varphi_{n+1}(t)$ obtains values in $[-q, q]$. Because

$$\varphi_n(t) = \sum_{k=1}^n (\varphi_n(t) \ominus \varphi_{n-1}(t)) \tag{6.72}$$

for any $\gamma < 1$, $t \in (-b, b)$ is selected in such a manner that $| \varphi_k(t) \ominus \varphi_{k-1}(t) | \leq \gamma^k$ for all k. This signifies that there exists $\gamma < 1$ [107] such that

$$| \varphi_k(t) \ominus \varphi_{k-1}(t) | \leq \gamma^k. \tag{6.73}$$

From the mean value theorem [160],

$$\varphi_k(t) \ominus \varphi_{k-1}(t) = \int_0^t [f(s, \varphi_{k-1}(s)) \ominus f(s, \varphi_{k-2}(s))]ds. \tag{6.74}$$

Applying the mean value theorem to the Z-number function $h(x) = f(s, x)$ at the two points $\varphi_{k-1}(s)$ and $\varphi_{k-2}(s)$,

$$h(\varphi_{k-1}(s)) \ominus h(\varphi_{k-2}(s)) = h'(\psi_k(s))(\varphi_{k-1}(s)) \ominus \varphi_{k-2}(s)). \tag{6.75}$$

Taking into consideration $h'(x) = \frac{\partial f}{\partial x}$, the following is obtained

$$\varphi_k(t) \ominus \varphi_{k-1}(t) = \int_0^t \frac{\partial f}{\partial x}(s, \psi_k(s))(\varphi_{k-1}(s) \ominus \varphi_{k-2}(s))ds. \tag{6.76}$$

Because $|\varphi_{k-1}(s) \ominus \varphi_{k-2}(s)| \leq \gamma^{k-1}$ for $s \leq t$ and $b < \gamma/N$, by substituting the above relation in (6.76) and utilizing the bounds of $\frac{\partial f}{\partial x}$,

$$|\varphi_k(t)) \ominus \varphi_{k-1}(t)| \leq \int_0^t N\gamma^{k-1}ds = Nt\gamma^{k-1} \leq Nb\gamma^{k-1}. \tag{6.77}$$

In order to validate that x is continuous, it is necessary to show that for any given $\epsilon > 0$ there exists $\delta > 0$ in such a manner that $|t_2 - t_1| < \delta$ implies $|\varphi(t_2) \ominus \varphi(t_1)| < \epsilon$. According to the notation convenience, it is supposed that $t_1 < t_2$. It follows that

$$\varphi(t_2) \ominus \varphi(t_1) = \lim_{n \to \infty} \varphi_n(t_2) \ominus \lim_{n \to \infty} \varphi_n(t_1)$$

$$= \lim_{n \to \infty} (\varphi_n(t_2) \ominus \varphi_n(t_1)) = \lim_{n \to \infty} \int_{t_1}^{t_2} f(s, \varphi_n(s))ds. \tag{6.78}$$

There exists N in such a manner that $|f(s, x)| \leq N$. Hence

$$|\varphi(t_2) \ominus \varphi(t_1)| \leq \int_{t_1}^{t_2} Nds = N|t_2 - t_1| \leq N\delta. \tag{6.79}$$

Therefore, by selecting $\delta < \epsilon/N$ it is observed that $|\varphi(t_2) \ominus \varphi(t_1)| < \epsilon$. So $\lim_{n \to \infty} \varphi_n(t)$ exists for all t.

Now it should be demonstrated that $\lim_{n \to \infty} \varphi_n(t)$ is continuous. Since

$$\varphi(t) = \lim_{n \to \infty} \varphi_n(t) = \lim_{n \to \infty} \int_0^t f(s, \varphi_{n-1}(s))ds$$

$$= \int_0^t \lim_{n \to \infty} f(s, \varphi_{n-1}(s))ds = \int_0^t f(s, \lim_{n \to \infty} \varphi_{n-1}(s))ds \tag{6.80}$$

where the last step (moving the limit inside the function) is at par with the concept that f is continuous in each variable. Hence it is clear that

$$\varphi(t) = \int_0^t f(s, \varphi(s))ds \tag{6.81}$$

because all functions are continuous,

$$\frac{\mathrm{d}}{\mathrm{d}t}\varphi = f(s,\varphi(t)). \tag{6.82}$$

If there exists another solution $\phi(t)$,

$$\varphi(t) \ominus \phi(t) = \int_0^t (f(s,\varphi(t)) \ominus f(s,\phi(t))) \mathrm{d}s. \tag{6.83}$$

Since the two functions are different, there exists $\epsilon > 0, |\varphi(t) \ominus \phi(t)| > \epsilon$. The following is defined

$$m = \max_{0 \le t \le b} |\varphi(t) \ominus \phi(t)|. \tag{6.84}$$

N is the bound for $\frac{\partial f}{\partial x}$. Utilizing the mean value theorem,

$$|\varphi(t) \ominus \phi(t)| \le \int_0^t N |\varphi(t) - \phi(t)| \, \mathrm{d}s \le N |t| m \le Nbm. \tag{6.85}$$

If $b < \epsilon/2mN$ is selected then it signifies that for all $t < b, |\varphi(t) - \phi(t)| < \epsilon/2$, indicating that the least difference is ϵ. So there exists a unique Z-number solution. ∎

Theorem 6.4 *Assume the following FDE based on Z-numbers*

$$\frac{\mathrm{d}}{\mathrm{d}t}x = f(t,x) \tag{6.86}$$

where $f \in \bar{J}_{ab}, \bar{J}_{ab}$ is the set of linear strongly bounded operators, for every operator f there exists a function $\tau \in L([a,b]; \tilde{Z}_+)$ such that $|f(v)(t)| \le \tau(t) \| v \|_G$, $t \in [a,b]$, and $v \in G([a,b]; \tilde{Z})$, and there exists $f_0, f_1 \in \varphi_{ab}$. φ_{ab} is a set of linear operators $f \in \bar{J}_{ab}$ from the set $G([a,b]; \tilde{Z}_+)$ to the set $L([a,b]; \tilde{Z}_+)$, such that

$$\begin{aligned}
|\underline{f}(t,\underline{v},\overline{v}) + \underline{f}_1(t,\underline{v},\overline{v})| &\le \underline{f}_0(t,|\underline{v}|,|\overline{v}|), \quad t \in [a,b] \\
|\overline{f}(t,\underline{v},\overline{v}) + \overline{f}_1(t,\underline{v},\overline{v})| &\le \overline{f}_0(t,|\underline{v}|,|\overline{v}|), \quad t \in [a,b]
\end{aligned} \tag{6.87}$$

then (6.86) has a unique solution.

Proof: If x is a Z-number solution of (6.86) and $-\frac{1}{2}f_1 \in J_{ab}(a)$,

$$\frac{\mathrm{d}}{\mathrm{d}t}\beta = -\frac{1}{2}f_1(t,\beta) \oplus f_0(t,|x|) \oplus \frac{1}{2}f_1(t,|x|) \tag{6.88}$$

has a unique Z-number solution β. Moreover as $f_0, f_1 \in \varphi_{ab}$

$$\begin{aligned}
\underline{\beta}(t) &\ge 0, \quad t \in [a,b] \\
\overline{\beta}(t) &\ge 0, \quad t \in [a,b]
\end{aligned}. \tag{6.89}$$

According to (6.87) and the condition $f_1 \in \varphi_{ab}$, from (6.88) the following is obtained

$$
\begin{aligned}
\frac{d}{dt}\underline{\beta} &\geq -\frac{1}{2}\underline{f}_{-1}(t,\underline{\beta},\overline{\beta}) + \underline{f}(t,\underline{x},\overline{x}) + \frac{1}{2}\underline{f}_{-1}(t,\underline{x},\overline{x}) \\
\frac{d}{dt}\overline{\beta} &\geq -\frac{1}{2}\overline{f}_1(t,\underline{\beta},\overline{\beta}) + \overline{f}(t,\underline{x},\overline{x}) + \frac{1}{2}\overline{f}_1(t,\underline{x},\overline{x})
\end{aligned}
\tag{6.90}
$$

thus $t \in [a,b]$

$$
\begin{aligned}
\frac{d}{dt}(-\underline{\beta}) &\leq -\frac{1}{2}\underline{f}_{-1}(t,-\underline{\beta},-\overline{\beta}) + \underline{k}(t,\underline{x},\overline{x}) + \frac{1}{2}\underline{k}_1(t,\underline{x},\overline{x}) \\
\frac{d}{dt}(-\overline{\beta}) &\leq -\frac{1}{2}\overline{f}_1(t,-\underline{\beta},-\overline{\beta}) + \overline{f}(t,\underline{x},\overline{x}) + \frac{1}{2}\overline{f}_1(t,\underline{x},\overline{x})
\end{aligned}
\tag{6.91}
$$

The last two inequalities are due to the assumption $-\frac{1}{2}f_1 \in J_{ab}(a)$

$$
\begin{aligned}
|\underline{x}(t)| &\leq \underline{\beta}(t) \quad t \in [a,b] \\
|\overline{x}(t)| &\leq \overline{\beta}(t) \quad t \in [a,b]
\end{aligned}
\tag{6.92}
$$

According to (6.92) and the conditions $f_0, f_1 \in \varphi_{ab}$, (6.88) results in

$$
\begin{aligned}
\frac{d}{dt}\underline{\beta} &\leq \underline{f}_{-0}(t,\underline{\beta},\overline{\beta}), \quad t \in [a,b] \\
\frac{d}{dt}\overline{\beta} &\leq f_0(t,\underline{\beta},\overline{\beta}), \quad t \in [a,b]
\end{aligned}
\tag{6.93}
$$

As $f_0 \in J_{ab}(a)$, the last inequality with $\beta(a) = 0$ gives $\underline{\beta}(t) \leq 0$ and $\overline{\beta}(t) \leq 0$ for $t \in [a,b]$. (6.89) implies $\beta \equiv 0$. Thus based on (6.92) it is concluded that $x \equiv 0$. ∎

6.7 Fuzzy Controller Design using Fuzzy Differential Equations and Z-number

In general, it is difficult to solve the four equations (4.6) or (4.5). Here, a special neural network named the Bernstein neural network is used to approximate the solutions of the FDE (4.5).

The Bernstein neural network uses the following Bernstein polynomial,

$$
B(x_1, x_2) = \sum_{i=0}^{N}\sum_{j=0}^{M} \binom{N}{i}\binom{M}{j}
\tag{6.94}
$$

$$
W_{i,j} x_{1i}(T - x_{1i})^{N-i} x_{2j}(1 - x_{2j})^{M-j}
$$

where $\binom{N}{i} = \frac{N!}{i!(N-i)!}$, $\binom{M}{j} = \frac{M!}{j!(M-j)!}$ and $W_{i,j}$ is the Z-number coefficient.

This two variable polynomial can be regarded as a neural network, which has two inputs x_{1i} and x_{2j} and one output y,

$$
y = \sum_{i=0}^{N}\sum_{j=0}^{M} \lambda_i \gamma_j W_{i,j} x_{1i}(T - x_{1i})^{N-i} x_{2j}(1 - x_{2j})^{M-j}
\tag{6.95}
$$

where $\lambda_i = \begin{pmatrix} N \\ i \end{pmatrix}$, $\gamma_j = \begin{pmatrix} M \\ j \end{pmatrix}$.

Because the Bernstein neural network (6.95) has similar forms as (4.6), the Bernstein neural network (6.95) is used to approximate the solutions of the four ordinary differential equations (ODEs) in (4.6).

If x_1 and x_2 in (6.94) are defined as: x_1 is time interval t, x_2 is the α-level, the solution of (4.5) in the form of the Bernstein neural network is

$$x_m(t, \alpha) = x_m(0, \alpha) \oplus t \sum_{i=0}^{N} \sum_{j=0}^{M} \lambda_i \gamma_j W_{i,j} t_i (T - t_i)^{N-i} \alpha_j (1 - \alpha_j)^{M-j} \tag{6.96}$$

where $x_m(0, \alpha)$ is the initial condition of the solution based on the Z-number.
So the derivative of (6.95) is

$$\begin{cases} (1) \begin{cases} \frac{d}{dt} \underline{x}_m = C_1 + C_2 \\ \frac{d}{dt} \overline{x}_m = D_1 + D_2 \end{cases} \\ (2) \begin{cases} \frac{d}{dt} \underline{x}_m = C_1 + C_2 \\ \frac{d}{dt} \overline{x}_m = D_1 + D_2 \end{cases} \end{cases} \tag{6.97}$$

where $t \in [0, T]$, $\alpha \in [0, 1]$, $t_k = kh$, $h = \frac{T}{k}$, $k = 1, ldots, N$, $\alpha_j = jh_1$, $h_1 = \frac{1}{M}$, $j = 1, \ldots, M$,

$$\begin{aligned} C_1 &= \sum_{i=0}^{N} \sum_{j=0}^{M} \lambda_i \gamma_j \underline{W}_{i,j} t_i (T - t_i)^{N-i} \alpha_j (1 - \alpha_j)^{M-j} \\ D_1 &= \sum_{i=0}^{N} \sum_{j=0}^{M} \lambda_i \gamma_j \overline{W}_{i,j} t_i (T - t_i)^{N-i} \alpha_j (1 - \alpha_j)^{M-j} \\ C_2 &= t_k \sum_{i=0}^{N} \sum_{j=0}^{M} \lambda_i \gamma_j \underline{W}_{i,j} [it_{i-1,j} (T - t_i)^{N-i} \\ &\quad - (N - i) t_{i,j} (T - t_i)^{N-i-1}] \alpha_j^i (1 - \alpha_j)^{M-j} \\ D_2 &= t_k \sum_{i=0}^{N} \sum_{j=0}^{M} \lambda_i \gamma_j \overline{W}_{i,j} [it_{i-1,j} (T - t_i)^{N-i} \\ &\quad - (N - i) t_{i,j} (T - t_i)^{N-i-1}] \alpha_j^i (1 - \alpha_j)^{M-j} \end{aligned} \tag{6.98}$$

The above equations can be considered as the neural network form, see Figure 4.1.

- Input unit

$$o_1^1 = t, \quad o_2^1 = \alpha. \tag{6.99}$$

- The first hidden units

$$\begin{aligned} o_{1,i}^2 &= f_i^1(o_1^1), \quad o_{2,i}^2 = f_i^2(o_1^1) \\ o_{3,j}^2 &= g_j^1(o_2^1), \quad o_{4,j}^2 = g_j^2(o_2^1) \end{aligned} \tag{6.100}$$

- The second hidden units

$$o_{1,i}^3 = o_{1,i}^2(o_{2,i}^2), \quad o_{2,j}^3 = o_{3,j}^2(o_{4,j}^2). \tag{6.101}$$

- The third hidden units

$$o_{1,i}^4 = \lambda_i o_{1,i}^3, \quad o_{2,i'}^4 = \gamma_j o_{2,j}^3. \tag{6.102}$$

- The fourth hidden units

$$o_{i,j}^5 = o_{1,i}^4 o_{2,j}^4. \tag{6.103}$$

- Output unit

$$N(t, \alpha) = \sum_{i=0}^{N} \sum_{j=0}^{M} (a_{i,j} o_{i,j}^5) \tag{6.104}$$

where $f_i^1 = t^i$, $f_i^2 = (T - t)^{N-i}$, $\lambda_i = \frac{1}{T^N}\begin{pmatrix} N \\ i \end{pmatrix}$, $g_j^1 = \alpha^j$, $g_j^2 = (1 - \alpha)^{M-j}$, $\gamma_j = \begin{pmatrix} M \\ j \end{pmatrix}$.

The training errors between (6.97) and (4.6) are defined as

$$(1) \begin{cases} \underline{e}_1 = C_1 + C_2 - \underline{f} \\ \overline{e}_1 = D_1 + D_2 - \overline{f} \end{cases}$$
$$(2) \begin{cases} \underline{e}_2 = C_1 + C_2 - \overline{f} \\ \overline{e}_2 = D_1 + D_2 - \underline{f} \end{cases}. \tag{6.105}$$

The standard back propagation learning algorithm is utilized to update the weights with the above training errors

$$\underline{W}_{i,j}(k + 1) = \underline{W}_{i,j}(k) - \eta_1 \left(\frac{\partial \underline{e}_1^2}{\partial \underline{W}_{i,j}} + \frac{\partial \overline{e}_1^2}{\partial \underline{W}_{i,j}} \right)$$
$$\overline{W}_{i,j}(k + 1) = \overline{W}_{i,j}(k) - \eta_2 \left(\frac{\partial \underline{e}_2^2}{\partial \overline{W}_{i,j}} + \frac{\partial \overline{e}_2^2}{\partial \overline{W}_{i,j}} \right) \tag{6.106}$$

where η_1 and η_2 are positive learning rates.

The momentum terms $\gamma \Delta \underline{W}_{i,j}(k - 1)$ and $\gamma \Delta \overline{W}_{i,j}(k - 1)$ can be added to stabilize the training process. The above Bernstein neural network can be converted to a recurrent (dynamic) form, see Figure 4.2. The dynamic Bernstein neural network is

$$\begin{cases} \frac{d}{dt}\underline{x}_m(t, \alpha) = \underline{P}(t, \alpha)A(\underline{x}_m(t, \alpha), \overline{x}_m(t, \alpha)) + \underline{Q}(t, \alpha) \\ \frac{d}{dt}\overline{x}_m(t, \alpha) = \overline{P}(t, \alpha)A(\underline{x}_m(t, \alpha), \overline{x}_m(t, \alpha)) + \overline{Q}(t, \alpha) \end{cases}. \tag{6.107}$$

Obviously, this dynamic network has the form of

$$f(t, x) = P(t)x + Q(t). \tag{6.108}$$

The training algorithm is similar to (6.106), only the training errors are changed as

$$
(1) \begin{cases} \underline{e}_1 = C_1 + C_2 - \underline{P}A(\underline{x}_m, \overline{x}_m) - \underline{Q} \\ \overline{e}_1 = D_1 + D_2 - \overline{P}A(\underline{x}_m, \overline{x}_m) - \overline{Q} \end{cases} \\
(2) \begin{cases} \underline{e}_2 = C_1 + C_2 - \overline{P}A(\underline{x}_m, \overline{x}_m) - \overline{Q} \\ \overline{e}_2 = D_1 + D_2 - \underline{P}A(\underline{x}_m, \overline{x}_m) - \underline{Q} \end{cases}
\tag{6.109}
$$

6.8 Approximation using a Fuzzy Sumudu Transform and Z-numbers

Consider the following fuzzy initial value problem based on Z-numbers

$$
\begin{cases} \dfrac{d}{dt}x(t) = f(t, x(t)), \\ x(0) = (\underline{x}(0, \alpha), \overline{x}(0, \alpha)), \quad 0 < \alpha \le 1 \end{cases}
\tag{6.110}
$$

where $f(t, x(t))$ is a Z-number function. The Z-number function $f(t, x(t))$ is the mapping of $f : R \times \tilde{Z} \to \tilde{Z}$. By utilizing the fuzzy Sumudu transform (FST) method for Z-numbers, the following is obtained

$$
S\left[\dfrac{d}{dt}x(t)\right] = S[f(t, x(t))]
\tag{6.111}
$$

The resolving process of Equation (6.111) is based on the following cases.

Case 1. Assume that $\dfrac{d}{dt}x(t)$ is (i)-differentiable. Based on Theorem 4.5 the following is extracted

$$
\dfrac{d}{dt}x(t) = \left(\dfrac{d}{dt}\underline{x}(t, \alpha), \dfrac{d}{dt}\overline{x}(t, \alpha)\right)
\tag{6.112}
$$

$$
S[\dfrac{d}{dt}x(t)] = \left(\dfrac{1}{B} \odot S[x(t)]\right) \ominus \dfrac{1}{B}x(0).
\tag{6.113}
$$

Equation (6.113) can be displayed as

$$
\begin{cases} S[\underline{f}(t, x(t), \alpha)] = \dfrac{1}{B}S[\underline{x}(t, \alpha)] - \dfrac{1}{B}\underline{x}(0, \alpha) \\ S[\overline{f}(t, x(t), \alpha)] = \dfrac{1}{B}S[\overline{x}(t, \alpha)] - \dfrac{1}{B}\overline{x}(0, \alpha) \end{cases}
\tag{6.114}
$$

where

$$
\begin{cases} \underline{f}(t, x(t), \alpha) = \min\{f(t, B)|B \in (\underline{x}(t, \alpha), \overline{x}(t, \alpha))\} \\ \overline{f}(t, x(t), \alpha) = \max\{f(t, B)|B \in (\underline{x}(t, \alpha), \overline{x}(t, \alpha))\} \end{cases}
\tag{6.115}
$$

Accordingly, Equation (6.115) can be solved on the basis of the following assumptions

$$
S[\underline{x}(t, \alpha)] = U_1(B, \alpha)
\tag{6.116}
$$

$$\mathbf{S}[\overline{x}(t, \alpha)] = U_2(B, \alpha) \tag{6.117}$$

where $U_1(B, \alpha)$ and $U_2(B, r)$ are the Z-number solutions of Equation (6.115). By applying the inverse Sumudu transform, $\underline{x}(t, \alpha)$ and $\overline{x}(t, \alpha)$ are computed as

$$\underline{x}(t, \alpha) = \mathbf{S}^{-1}[U_1(B, \alpha)] \tag{6.118}$$

$$\overline{x}(t, \alpha) = \mathbf{S}^{-1}[U_2(B, \alpha)]. \tag{6.119}$$

Case 2. Assume that $\frac{d}{dt}x(t)$ is (ii)-differentiable. Based on Theorem 4.5 the following is extracted

$$\frac{d}{dt}x(t) = \left(\frac{d}{dt}\overline{x}(t, \alpha), \frac{d}{dt}\underline{x}(t, \alpha) \right) \tag{6.120}$$

$$\mathbf{S}\left[\frac{d}{dt}x(t) \right] = \left(\frac{-1}{B} \odot x(0) \right) \ominus \left(\frac{-1}{B} \odot \mathbf{S}[x(t)] \right). \tag{6.121}$$

Equation (6.121) can be displayed as

$$\begin{cases} \mathbf{S}[\underline{f}(t, x(t), \alpha)] = \frac{1}{B}\mathbf{S}[\underline{x}(t, \alpha)] - \frac{1}{B}\underline{x}(0, \alpha) \\ \mathbf{S}[\overline{f}(t, x(t), \alpha)] = \frac{1}{B}\mathbf{S}[\overline{x}(t, \alpha)] - \frac{1}{B}\overline{x}(0, \alpha) \end{cases} \tag{6.122}$$

where

$$\begin{cases} \underline{f}(t, x(t), \alpha) = \min\{f(t, B)|B \in (\underline{x}(t, \alpha), \overline{x}(t, \alpha))\} \\ \overline{f}(t, x(t), \alpha) = \max\{f(t, B)|B \in (\underline{x}(t, \alpha), \overline{x}(t, \alpha))\} \end{cases}. \tag{6.123}$$

Accordingly, Equation (6.123) can be resolved on the basis of the following assumptions

$$\begin{aligned} \mathbf{S}(\underline{x}(t, \alpha) &= V_1(B, \alpha) \\ \mathbf{S}(\overline{x}(t, \alpha) &= V_2(B, \alpha) \end{aligned} \tag{6.124}$$

where $V_1(B, \alpha)$ and $V_2(B, \alpha)$ are the Z-number solutions of the Equation (6.123). By applying the inverse Sumudu transform, $\underline{x}(t, \alpha)$ and $\overline{x}(t, \alpha)$ are computed as

$$\begin{aligned} \underline{x}(t, \alpha) &= \mathbf{S}^{-1}[V_1(B, \alpha)] \\ \overline{x}(t, \alpha) &= \mathbf{S}^{-1}[V_2(B, \alpha)] \end{aligned}. \tag{6.125}$$

6.9 Simulations

In this section, several real applications are used to show the use of the fuzzy equation and the FDE with Z-number coefficients to design the fuzzy controller.

Example 6.1 Chemical reaction In the chemical reaction the poly ethylene (PE) and poly propylene (PP) are used to generate a desired substance (DS). The

Table 6.1 Neural networks approximate the *Z*-numbers.

k	x (k) NN	k	x (k) FNN
1	$[(22.53, 23.68, 24.10), p(0.6, 0.8, 0.85)]$	1	$[(22.32, 23.48, 23.97), p(0.7, 0.8, 0.85)]$
2	$[(21.793, 22.837, 23.203), p(0.7, 0.8, 0.85)]$	2	$[(20.98, 22.133, 22.76), p(0.7, 0.85, 0.9)]$
⋮	⋮	⋮	⋮
35	$[(18.67, 19.71, 20.23), p(0.8, 0.92, 1)]$	18	$[(18.49, 19.51, 20.13), p(0.8, 0.92, 1)]$
36	$[(18.38, 19.40, 19.91), p(0.8, 0.96, 1)]$	19	$[(18.37, 19.39, 19.90), p(0.8, 0.96, 1)]$

cost of the material is defined as x, the cost of PE is x and the cost of PP is x^2. The weights of PE and PP are uncertain, and satisfy the triangle function (6.1). It is a requirement to produce two types of DS. If the cost between them is desired to be $[(360.5009, 400.5565, 421.3749), p(0.8, 0.9, 1)] = y^*$, what is the cost of x? The weights of PE are

$$a_1 = [(2.7951, 3.35412, 3.9131), p(0.7, 0.8, 1)] \\ b_1 = [(1.5811, 2.1081, 2.6352), p(0.8, 0.9, 1)]$$ (6.126)

The PP weights are

$$a_2 = [(4.8107, 5.3452, 5.8797), p(0.7, 0.875, 1)] \\ b_2 = [(3.9131, 4.4721, 5.0311), p(0.6, 0.8, 1)]$$ (6.127)

The reaction can be modeled with the following fuzzy equation and *Z*-numbers

$$[(2.7951, 3.35412, 3.9131), p(0.7, 0.8, 1)]x \\ \oplus [(4.8107, 5.3452, 5.8797), p(0.7, 0.875, 1)]x^2 \\ = [(1.5811, 2.1081, 2.6352), p(0.8, 0.9, 1)]x \\ \oplus [(3.9131, 4.4721, 5.0311), p(0.6, 0.8, 1)]x^2 \\ \oplus [(360.5009, 400.5565, 421.3749), p(0.8, 0.9, 1)]$$ (6.128)

where $f_1(x) = g_1(x) = x$ and $f_2(x) = g_2(x) = x^2$.

The exact solution is $x^* = [(18.3712, 19.3919, 19.9022), p(0.8, 0.96, 1)]$. Feed forward neural networks (NNs), as in Figure 3.2, and feed back neural networks (FNNs), as Figure 3.3, are used to approximate the solution x. The learning rate is $\eta = 0.02$. The initial state is

$$x(0) = [(22.6612, 23.7102, 24.2407), p(0.8, 0.9, 1)].$$

The approximation results are shown in Table 6.1. The modeling errors are shown in Figure 6.2.

It can be seen that both the neural networks work well. The following relation is utilized to transfer the *Z*-numbers to fuzzy numbers,

$$\alpha = \frac{\int x \pi_{\tilde{p}}(x) dx}{\int \pi_{\tilde{p}}(x) dx}$$ (6.129)

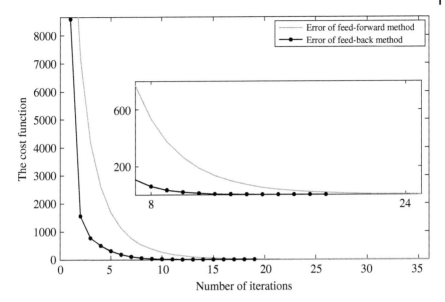

Figure 6.2 Approximation errors of the neural networks.

Table 6.2 Neural networks approximate the fuzzy numbers.

k	$x(k)$ NN	k	$x(k)$FNN
1	(19.358, 20.349, 20.709)	1	(19.672, 20.698, 21.132)
2	(19.205, 20.125, 20.448)	2	(18.844, 19.878, 20.442)
⋮	⋮	⋮	⋮
35	(17.720, 18.710, 19.203)	18	(17.548, 18.517, 19.107)
36	(17.541, 18.513, 19.000)	19	(17.538, 18.509, 18.996)

$Z = (\tilde{A}, \tilde{P}) = [(22.331, 23.384, 23.993), p(0.6, 0.8, 0.85)]$. So

$$\tilde{Z}^{\alpha} = (22.331, 23.384, 23.993; 0.7)$$

$\tilde{Z}' = (\sqrt{0.7}22.331, \sqrt{0.7}23.384, \sqrt{0.7}23.993)$. The results of the neural network approximation for the fuzzy numbers are shown in Table 6.2.

The Z-numbers increase the degree of reliability of the information. The comparison between the Z-number $Z = [(18.382, 19.401, 19.911), p(0.8, 0.96, 1)]$ and the fuzzy number (17.541, 18.513, 19.000) for $k = 36$ is shown in Figure 6.3. It can be seen that the Z-number incorporates a variety of information and the solution of the Z-number is more accurate. The membership function for

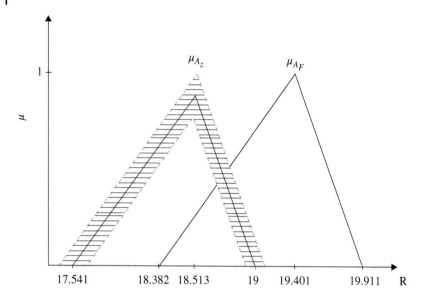

Figure 6.3 Z-number and fuzzy number.

the restriction in the Z-number is $\mu_{A_Z} = (18.382, 19.401, 19.911)$. It can be in probability form. ∎

Example 6.2 Heat source The heat source is in the center of insulating materials, see Figure 3.5. The thickness of the materials is not exact, and satisfy the trapezoidal function (6.2),

$$
\begin{aligned}
A &= [(0.1317, 0.1536, 0.1646, 0.1975), p(0.7, 0.83, 0.9)] = a_1 \\
B &= [(0.0843, 0.1054, 0.2108, 0.5270), p(0.8, 0.9, 1)] = a_2 \\
C &= [(0.0964, 0.1072, 0.2144, 0.4288), p(0.7, 0.87, 0.9)] = b_1 \\
D &= [(0.0216, 0.0325, 0.0542, 0.0867), p(0.8, 0.85, 0.92)] = b_2
\end{aligned} \qquad (6.130)
$$

The coefficient associated with the conductivity materials are $K_A = x = f_1$, $K_B = x\sqrt{x} = f_2$, $K_C = x^2 = g_1$, and $K_D = \sqrt{x} = g_2$, where x is considered to be the elapsed time. The control objective is to find the time when the thermal resistance at the right side arrives at $R = [(0.0162, 0.0293, 0.0424, 0.1241), p(0.75, 0.8, 0.9)] = y^*$. The thermal balance model is [81]:

$$
\frac{A}{K_A} \oplus \frac{B}{K_B} = \frac{C}{K_C} \oplus \frac{D}{K_D} \oplus R. \qquad (6.131)
$$

The exact solution is $x^* = [(2.0519, 3.0779, 4.1039, 6.1559), p(0.8, 0.95, 1)]$ [81]. The learning rate is $\eta = 0.1$ (NN) and $\eta = 0.005$ (FNN). The neural networks approximation results are shown in Tables 6.3 and 6.4. ∎

Table 6.3 Neural networks approximate the Z-numbers of the heat source.

k	$x(k)$NN
1	$[(5.972, 6.983, 7.963, 9.982), p(0.6, 0.8, 0.85)]$
2	$[(5.438, 6.383, 7.353, 9.302), p(0.75, 0.8, 0.9)]$
$\vdots$	$\vdots$
53	$[(2.118, 3.170, 4.224, 6.333), p(0.8, 0.9, 1)]$
54	$[(2.069, 3.087, 4.113, 6.175), p(0.8, 0.94, 1)]$

k	$x(k)$FNN
1	$[(5.989, 6.990, 7.979, 9.988), p(0.7, 0.85, 0.87)]$
2	$[(5.378, 6.102, 7.123, 9.162), p(0.7, 0.85, 0.87)]$
$\vdots$	$\vdots$
22	$[(2.089, 3.149, 4.146, 6.292), p(0.8, 0.96, 1)]$
23	$[(2.059, 3.084, 4.113, 6.169), p(0.8, 0.94, 1)]$

Table 6.4 Neural networks approximate the fuzzy numbers of the heat source.

k	$x(k)$ with NN	k	$x(k)$ with FNN
1	$(5.131, 5.999, 6.841, 8.576)$	1	$(5.360, 6.255, 7.141, 8.939)$
2	$(4.934, 5.791, 6.671, 8.440)$	2	$(4.813, 5.461, 6.374, 8.199)$
$\vdots$	$\vdots$	$\vdots$	$\vdots$
53	$(2.009, 3.007, 4.007, 6.008)$	22	$(1.993, 3.004, 3.956, 6.004)$
54	$(1.966, 2.934, 3.915, 5.870)$	23	$(1.957, 2.931, 3.909, 5.864)$

Example 6.3 Water channel system The water in the pipe d_1 is divided into three pipes d_2, d_3, d_4, see Figure 3.8. The areas of the pipes are uncertain, they satisfy the trapezoidal function (6.2),

$$\begin{aligned} A_1 &= [(0.421, 0.632, 0.737, 0.843), p(0.75, 0.9, 1)] \\ A_2 &= [(0.052, 0.104, 0.209, 0.419), p(0.8, 0.91, 1)] \\ A_3 &= [(0.031, 0.084, 0.105, 0.210), p(0.8, 0.9, 0.95)] \end{aligned} \qquad (6.132)$$

The water velocities in the pipes are controlled by the valves parameter x, $v_1 = x^3$, $v_2 = \frac{e^x}{2}$, $v_3 = x$ [177]. The control objective is to allow the flow in pipe d_4, which is

$$Q = [(11.478, 40.890, 93.332, 293.056), p(0.8, 0.87, 0.95)]. \qquad (6.133)$$

What is the valve control parameter x? By mass balance

$$A_1 v_1 = A_2 v_2 \oplus A_3 v_3 \oplus Q. \qquad (6.134)$$

Table 6.5 Neural networks approximate the Z-numbers of the water channel system.

k	x (k) NN
1	$[(5.751, 6.772, 7.741, 9.761), p(0.6, 0.8, 0.85)]$
2	$[(5.327, 6.261, 7.131, 9.201), p(0.7, 0.8, 0.87)]$
⋮	⋮
55	$[(3.142, 4.194, 5.229, 7.321), p(0.8, 0.9, 1)]$
56	$[(3.136, 4.188, 5.222, 7.315), p(0.8, 0.93, 1)]$

k	x (k) FNN
1	$[(5.878, 6.881, 7.867, 9.877), p(0.7, 0.81, 0.85)]$
2	$[(5.158, 6.004, 7.001, 9.002), p(0.7, 0.85, 0.9)]$
⋮	⋮
20	$[(3.136, 4.185, 5.225, 7.312), p(0.85, 0.9, 1)]$
21	$[(3.130, 4.178, 5.218, 7.305), p(0.8, 0.92, 1)]$

Table 6.6 Neural networks approximate the fuzzy numbers of the water channel system.

k	x (k) NN	k	x (k) FNN
1	$(4.941, 5.818, 6.651, 8.386)$	1	$(5.188, 6.073, 6.944, 8.718)$
2	$(4.720, 5.548, 6.319, 8.153)$	2	$(4.632, 5.392, 6.287, 8.085)$
⋮	⋮	⋮	⋮
55	$(2.980, 3.978, 4.960, 6.945)$	20	$(2.975, 3.970, 4.956, 6.936)$
56	$(2.977, 3.976, 4.958, 6.945)$	21	$(2.970, 3.964, 4.951, 6.932)$

The exact solution is demonstrated by $x = [(3.127, 4.170, 5.212, 7.298), p(0.8, 0.92, 1)]$ [177]. The learning rate of NN is $\eta = 0.08$. The neural networks approximation results are shown in Tables 6.5 and 6.6.

FNN is much faster and more robust compared with NN. After converting the Z-numbers to fuzzy numbers, it is possible to obtain the fuzzy rules. ∎

Example 6.4 Heat treatment system The heat treatment system in welding can be modeled as [55]:

$$\frac{\mathrm{d}}{\mathrm{d}t}x(t) = 3Ax^2(t) \tag{6.135}$$

where transfer area A is uncertain as $A = [(1 + \alpha, 3 - \alpha), p(0.8, 0.87, 0.95)]$, $\alpha \in [0, 1]$. So (6.135) is an FDE based on the Z-number. If the initial condition

Table 6.7 Bernstein neural networks approximate the *Z*-numbers of the heat treatment system.

α	SNN	DNN
0	[(0.0582,0.0859),p(0.7,0.8,0.85)]	[(0.0250,0.0425),p(0.7,0.82,0.9)]
0.1	[(0.0449,0.0696),p(0.7,0.8,0.9)]	[(0.0224,0.0399),p(0.75,0.82,0.9)]
0.2	[(0.0419,0.0619),p(0.8,0.92,1)]	[(0.0207,0.0394),p(0.8,0.94,1)]
0.3	[(0.0250,0.0348),p(0.7,0.81,0.9)]	[(0.0226,0.0344),p(0.8,0.85,0.96)]
0.4	[(0.0487,0.0689),p(0.7,0.8,0.88)]	[(0.0271,0.0510),p(0.75,0.82,0.9)]
0.5	[(0.0534,0.0665),p(0.8,0.9,1)]	[(0.0160,0.0271),p(0.81,0.92,1)]
0.6	[(0.0494,0.0765),p(0.8,0.9,1)]	[(0.0201,0.0413),p(0.81,0.92,1)]
0.7	[(0.0630,0.0859),p(0.75,0.82,0.9)]	[(0.0303,0.0476),p(0.82,0.9,1)]
0.8	[(0.0393,0.0536),p(0.8,0.92,1)]	[(0.0164,0.0379),p(0.82,0.94,1)]
0.9	[(0.0422,0.0669),p(0.8,0.9,1)]	[(0.0212,0.0430),p(0.8,0.94,1)]
1	[(0.0443,0.0443),p(0.7,0.8,0.88)]	[(0.0186,0.0186),p(0.7,0.82,0.9)]

is $x(0) = [(0.5\sqrt{\alpha}, 0.2\sqrt{1-\alpha}+0.5), p(0.8, 0.92, 1)]$, the static Bernstein neural network (4.68) has the form of

$$
\begin{cases}
\underline{x}_m(t, \alpha) = 0.5\sqrt{\alpha} \\
\quad + t \sum_{i=0}^{N} \sum_{j=0}^{M} \lambda_i \gamma_j \underline{W}_{i,j} t_i (T - t_i)^{N-i} \alpha_j (1 - \alpha_j)^{M-j} \\
\overline{x}_m(t, \alpha) = 0.2\sqrt{1-\alpha} + 0.5 \\
\quad + t \sum_{i=0}^{N} \sum_{j=0}^{M} \lambda_i \gamma_j \overline{W}_{i,j} t_i (T - t_i)^{N-i} \alpha_j (1 - \alpha_j)^{M-j}
\end{cases}
\tag{6.136}
$$

where the approximate *Z*-number solution is termed as

$$[(\underline{x}_m(t, \alpha), \overline{x}_m(t, \alpha)), p(0.8, 0.9, 1)].$$

With the learning rates $\eta = 0.002$ and $\gamma = 0.002$, the approximation results for *Z*-numbers are shown in Table 6.7. The results of the Bernstein neural network approximation for the fuzzy numbers are shown in Table 6.8. ∎

Example 6.5 Tank system   A generalized model of a tank system is displayed in Figure 4.4. Assume $I = t + 1$ to be inflow disturbances of the tank that will generate vibration in the liquid level x; here $R = 1$ is the flow obstruction that can be curbed using the valve and $A = 1$ is considered to be the cross section of the mentioned tank. The expression in relation to the liquid level considering the tank can be described as [177]:

$$\frac{d}{dt}x(t) = -\frac{1}{AR}x(t) + \frac{I}{A}. \tag{6.137}$$

Table 6.8 Bernstein neural networks approximate the fuzzy numbers of the heat treatment system.

α	SNN	DNN
0	[0.0511,0.0754]	[0.0224,0.0381]
0.1	[0.0402,0.0623]	[0.0203,0.0362]
0.2	[0.0398,0.0588]	[0.0197,0.0374]
0.3	[0.0224,0.0312]	[0.0211,0.0321]
0.4	[0.0433,0.0613]	[0.0246,0.0462]
0.5	[0.0507,0.0631]	[0.0152,0.0258]
0.6	[0.0469,0.0726]	[0.0191,0.0392]
0.7	[0.0571,0.0778]	[0.0288,0.0452]
0.8	[0.0373,0.0509]	[0.0157,0.0362]
0.9	[0.0401,0.0635]	[0.0202,0.0408]
1	[0.0394,0.0394]	[0.0167,0.0167]

If the initial condition is $x(0) = [(0.96 + 0.04\alpha, 1.01 - 0.01\alpha), p(0.75, 0.82, 0.9)]$, the static Bernstein neural network (4.68) has the form of

$$\begin{cases} \underline{x}_m(t, \alpha) = (0.96 + 0.04\alpha) \\ \qquad + t \sum_{i=0}^{N} \sum_{j=0}^{M} \lambda_i \gamma_j \underline{W}_{i,j} t_i (T - t_i)^{N-i} \alpha_j (1 - \alpha_j)^{M-j} \\ \overline{x}_m(t, \alpha) = (1.01 - 0.01\alpha) \\ \qquad + t \sum_{i=0}^{N} \sum_{j=0}^{M} \lambda_i \gamma_j \overline{W}_{i,j} t_i (T - t_i)^{N-i} \alpha_j (1 - \alpha_j)^{M-j} \end{cases} \tag{6.138}$$

where $t \in [0, 1]$ and the approximate Z-number solution is termed as

$$[(\underline{x}_m(t, \alpha), \overline{x}_m(t, \alpha)), p(0.75, 0.81, 0.95)].$$

Also, the dynamic Bernstein neural network (4.79) is utilized to approximate the solutions. To compare the results, the other generalization of a neural network method [67] is used. The comparison results for Z-numbers are shown in Table 6.9. The specifications quoted here are $\eta = 0.001$ and $\gamma = 0.001$. The results of the Bernstein neural network approximation for the fuzzy numbers are shown in Table 6.10. ∎

Example 6.6 Vibration mass system The vibration mass system shown in Figure 1.3 can be modeled by the ODE

$$\frac{\mathrm{d}}{\mathrm{d}t} x(t) = \frac{k}{m} x(t) \tag{6.139}$$

Table 6.9 Solutions of different methods based on the Z-numbers of the tank system.

α	SNN	DNN
0	[(0.0435, 0.0994),p(0.72,0.81,0.87)]	[(0.0112, 0.0442),p(0.75,0.82,0.88)]
0.2	[(0.0504, 0.0940),p(0.7,0.8,0.9)]	[(0.0248, 0.0635),p(0.75,0.82,0.9)]
0.4	[(0.0441, 0.0802),p(0.8,0.85,0.92)]	[(0.0131, 0.0422),p(0.8,0.9,1)]
0.6	[(0.0178, 0.0423),p(0.8,0.92,1)]	[(0.0121, 0.0384),p(0.81,0.94,1)]
0.8	[(0.0608, 0.0709),p(0.71,0.8,0.9)]	[(0.0154, 0.0309),p(0.8,0.87,0.95)]
1	[(0.0611, 0.0611),p(0.75,0.82,0.91)]	[(0.0335, 0.0335),p(0.8,0.87,0.92)]
α	Neural network	
0	[(0.0798, 0.1153),p(0.7,0.75,0.85)]	
0.2	[(0.0878, 0.1375),p(0.7,0.8,0.85)]	
0.4	[(0.1105, 0.1592),p(0.75,0.83,0.9)]	
0.6	[(0.0613, 0.0915),p(0.8,0.9,1)]	
0.8	[(0.0739, 0.0925),p(0.7,0.8,0.85)]	
1	[(0.1007, 0.1007),p(0.7,0.8,0.9)]	

Table 6.10 Solutions of different methods based on the fuzzy numbers of the tank system.

α	SNN	DNN	Neural network
0	[0.0387, 0.0884]	[0.0101, 0.0398]	[0.0701, 0.1012]
0.2	[0.0451, 0.0841]	[0.0225, 0.0575]	[0.0771, 0.1207]
0.4	[0.0407, 0.0740]	[0.0125, 0.0401]	[0.1001, 0.1442]
0.6	[0.0169, 0.0402]	[0.0115, 0.0365]	[0.0582, 0.0868]
0.8	[0.0544, 0.0635]	[0.0144, 0.0289]	[0.0649, 0.0812]
1	[0.0554, 0.0554]	[0.0311, 0.0311]	[0.0901, 0.0901]

where the spring constant is $k = 1$. The mass m is changeable in

$$[(0.75, 1.125), p(0.7, 0.8, 1)]$$

so the position state $x(t)$ has some uncertainties and the ODE (6.139) can be formed into an FDE based on the Z-number. It has the same form as (6.139), only $x(t)$ becomes a Z-number variable. If the initial position is $x(0) = [(0.75 + 0.25\alpha, 1.125 - 0.125\alpha), p(0.8, 0.9, 1)]$, $\alpha \in [0, 1]$, then the exact solutions of the FDE (6.139) is [78]

$$x(t, \alpha) = \left[((0.75 + 0.25\alpha)e^t, (1.125 - 0.125\alpha)e^t), p(0.8, 0.9, 1)\right] \quad (6.140)$$

where $t \in [0, 1]$. Now the static Bernstein neural network (4.68) is used to approximate the Z-number solution $[(\underline{x}_m(t, \alpha), \overline{x}_m(t, \alpha)), p(0.8, 0.94, 1)]$ (6.140)

Table 6.11 Solutions of different methods based on the *Z*-numbers of the vibration mass system.

α	Exact solution	SNN
0	[(2.1858,3.2787),p(0.8,0.87,0.95)]	[(2.2967,3.4240),p(0.7,0.81,0.85)]
0.2	[(2.2924,3.1521),p(0.81,0.9,1)]	[(2.3545,3.2570),p(0.7,0.82,0.9)]
0.6	[(2.5790,3.0088),p(0.81,0.9,1)]	[(2.6759,3.1461),p(0.7,0.8,0.87)]
1	[(2.9144,2.9144),p(0.8,0.87,0.95)]	[(2.9667,2.9667),p(0.7,0.8,0.87)]

α	Exact solution	Max-min Euler
0	[(2.1858,3.2787),p(0.8,0.87,0.95)]	[(2.4847,3.4771),p(0.7,0.82,0.85)]
0.2	[(2.2924,3.1521),p(0.81,0.9,1)]	[(2.6100,3.5888),p(0.72,0.8,0.87)]
0.6	[(2.5790,3.0088),p(0.81,0.9,1)]	[(2.7137,3.1660),p(0.6,0.8,0.87)]
1	[(2.9144,2.9144),p(0.8,0.87,0.95)]	[(3.0152,3.0152),p(0.6,0.8,0.87)]

α	DNN	Average Euler
0	[(2.2250,3.3883),p(0.71,0.85,0.87)]	[(2.9921,3.4921),p(0.65,0.8,0.85)]
0.2	[(2.3504,3.2467),p(0.75,0.83,0.9)]	[(2.8137,3.2303),p(0.6,0.7,0.75)]
0.6	[(2.6097,3.0872),p(0.75,0.83,0.9)]	[(2.9565,3.1372),p(0.6,0.7,0.8)]
1	[(2.9532,2.9532),p(0.71,0.85,0.87)]	[(3.1249,3.1249),p(0.6,0.7,0.8)]

where

$$
\begin{cases}
\underline{x}_m(t,\alpha) = (0.75 + 0.25\alpha) \\
\quad + t \sum_{i=0}^{N} \sum_{j=0}^{M} \lambda_i \gamma_j \underline{W}_{i,j} t_i (T - t_i)^{N-i} \alpha_j (1 - \alpha_j)^{M-j} \\
\overline{x}_m(t,\alpha) = (1.125 - 0.125\alpha) \\
\quad + t \sum_{i=0}^{N} \sum_{j=0}^{M} \lambda_i \gamma_j \overline{W}_{i,j} t_i (T - t_i)^{N-i} \alpha_j (1 - \alpha_j)^{M-j}
\end{cases}
\tag{6.141}
$$

The dynamic Bernstein neural network (4.79) is also utilized to approximate the solutions. The learning rates are $\eta = 0.01$, $\gamma = 0.01$. To compare the results, two popular methods are used: the max-min Euler method and average Euler method [185]. The comparison results are shown in Tables 6.11 and 6.12. Corresponding solution plots are shown in Figure 6.4. ∎

Example 6.7 Heating system A tank with a heating system is displayed in Figure 2.3, where $R = 0.5$, the thermal capacitance is $C = 2$, and the temperature is x. The model is formulated as (2.3). By utilizing the FST method based on the *Z*-number, the following results are obtained

$$
\mathbf{S} \left[\frac{\mathrm{d}}{\mathrm{d}t} x(t) \right] = \mathbf{S}[-x(t)]
\tag{6.142}
$$

Table 6.12 Approximation errors based on the Z-numbers of the vibration mass system.

α	SNN	DNN
0	[(0.0684,0.1251),p(0.7,0.8,0.85)]	[(0.0231,0.0671),p(0.7,0.85,0.87)]
0.2	[(0.0735,0.1192),p(0.7,0.8,0.9)]	[(0.0266,0.0675),p(0.75,0.8,0.9)]
0.6	[(0.0855,0.1095),p(0.8,0.87,0.95)]	[(0.0339,0.0689),p(0.8,0.9,1)]
0.8	[(0.0833,0.0939),p(0.8,0.91,1)]	[(0.0345,0.0526),p(0.8,0.94,1)]
1	[(0.1029,0.1029),p(0.7,0.8,0.9)]	[(0.0572,0.0572),p(0.8,0.85,0.95)]

α	Max-min Euler	Average Euler
0	[(0.1064,0.1596),p(0.7,0.8,0.85)]	[(0.2404,0.5138),p(0.6,0.8,0.85)]
0.2	[(0.1127,0.1551),p(0.7,0.8,0.87)]	[(0.1588,0.4286),p(0.7,0.8,0.85)]
0.6	[(0.1253,0.1462),p(0.7,0.85,0.9)]	[(0.0082,0.2798),p(0.7,0.81,0.9)]
0.8	[(0.1247,0.1345),p(0.8,0.9,1)]	[(0.0628,0.2009),p(0.75,0.9,1)]
1	[(0.1410,0.1410),p(0.7,0.8,0.87)]	[(0.1410,0.1410),p(0.7,0.8,0.87)]

$$\mathbf{S}\left[\frac{d}{dt}x(t)\right] = \int_0^\infty \frac{d}{dt}x(Bt) \odot e^{-t}dt \qquad (6.143)$$

where $0 \leq B < K$. If $x(t)$ is (i)-differentiable and case 1 holds, then

$$\mathbf{S}\left[\frac{d}{dt}x(t)\right] = \frac{1}{B} \odot (\mathbf{S}[x(t)] \ominus x(0)) = \frac{1}{B}\mathbf{S}[x(t)] \ominus \frac{1}{B}x(0). \qquad (6.144)$$

Therefore

$$-\mathbf{S}[x(t)] = \frac{1}{B}\mathbf{S}[x(t)] \ominus \frac{1}{B}x(0). \qquad (6.145)$$

Based on Equation (6.114), the following is obtained

$$\begin{cases} -\mathbf{S}[\overline{x}(t,\alpha)] = \frac{1}{B}\mathbf{S}[\underline{x}(t,\alpha)] - \frac{1}{B}\underline{x}(0,\alpha) \\ -\mathbf{S}[\underline{x}(t,\alpha)] = \frac{1}{B}\mathbf{S}[\overline{x}(t,\alpha)] - \frac{1}{B}\overline{x}(0,\alpha) \end{cases}. \qquad (6.146)$$

Therefore, the Z-number solution of Equation (6.146) is as follows

$$\begin{cases} \mathbf{S}[\overline{x}(t,\alpha)] = \left(\frac{-1}{B^2-1}\right)\overline{x}(0,\alpha) + \left(\frac{B}{B^2-1}\right)\underline{x}(0,\alpha) \\ \mathbf{S}[\underline{x}(t,\alpha)] = \left(\frac{-1}{B^2-1}\right)\underline{x}(t,\alpha) + \left(\frac{B}{B^2-1}\right)\overline{x}(0,\alpha) \end{cases}. \qquad (6.147)$$

By utilizing the inverse Sumudu transform for Z-numbers, the following is obtained

$$\begin{cases} \mathbf{S}[\overline{x}(t,\alpha)] = \overline{x}(0,\alpha)\mathbf{S}^{-1}\left(\frac{-1}{B^2-1}\right) + \underline{x}(0,\alpha)\mathbf{S}^{-1}\left(\frac{B}{B^2-1}\right) \\ \mathbf{S}[\underline{x}(t,\alpha)] = \underline{x}(0,\alpha)\mathbf{S}^{-1}\left(\frac{-1}{B^2-1}\right) + \overline{x}(0,\alpha)\mathbf{S}^{-1}\left(\frac{B}{B^2-1}\right) \end{cases} \qquad (6.148)$$

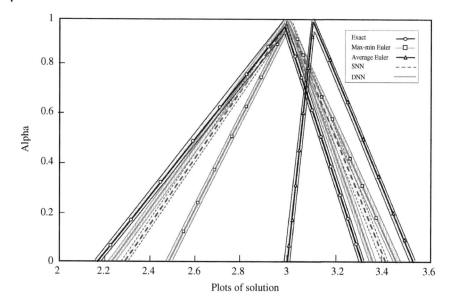

Figure 6.4 Comparison plots of the SNN, DNN, max-min Euler, and the average Euler methods, and the exact solution based on Z-numbers.

where

$$\begin{cases} \overline{x}(t,\alpha) = e^t \left(\frac{\overline{x}(0,\alpha)-\underline{x}(0,\alpha)}{2} \right) + e^{-t} \left(\frac{\overline{x}(0,\alpha)+\underline{x}(0,\alpha)}{2} \right) \\ \underline{x}(t,\alpha) = e^t \left(\frac{\underline{x}(0,\alpha)-\overline{x}(0,\alpha)}{2} \right) + e^{-t} \left(\frac{\underline{x}(0,\alpha)+\overline{x}(0,\alpha)}{2} \right) \end{cases}. \tag{6.149}$$

Now if $x(t)$ is (ii)-differentiable and case 2 holds, then

$$\mathbf{S}\left[\frac{d}{dt}x(t) \right] = \left(\frac{-1}{B}\mathbf{S}[x(t)] \right) \ominus \left(\frac{-1}{B}x(0) \right). \tag{6.150}$$

Hence

$$-\mathbf{S}[x(t)] = \left(\frac{-1}{B}\mathbf{S}[x(t)] \right) \ominus \left(\frac{-1}{B}x(0) \right). \tag{6.151}$$

Based on the above relations, Equation (2.3) can be written as follows

$$\begin{cases} -\mathbf{S}[\underline{x}(t,\alpha)] = \frac{1}{B}\mathbf{S}[\underline{x}(t,\alpha)] - \frac{1}{B}\underline{x}(0,\alpha) \\ -\mathbf{S}[\overline{x}(t,\alpha)] = \frac{1}{B}\mathbf{S}[\overline{x}(t,\alpha)] - \frac{1}{B}\overline{x}(0,\alpha) \end{cases}. \tag{6.152}$$

So, the Z-number solution of Equation (6.152) is displayed as

$$\begin{cases} \mathbf{S}[\underline{x}(t,\alpha)] = \underline{x}(0,\alpha) \left(\frac{1}{B+1} \right) \\ \mathbf{S}[\overline{x}(t,\alpha)] = \overline{x}(t,\alpha) \left(\frac{1}{B+1} \right) \end{cases}. \tag{6.153}$$

Table 6.13 Approximation errors based on the Z-numbers of the heating system.

α	FST	Average Euler
0	[(0.0078,0.0195),p(0.8,0.86,0.94)]	[(0.0138,0.0215),p(0.7,0.8,0.87)]
0.2	[(0.0085,0.0169),p(0.75,0.8,0.9)]	[(0.0188,0.0286),p(0.7,0.8,0.87)]
0.6	[(0.0058,0.0115),p(0.8,0.9,1)]	[(0.0182,0.0198),p(0.7,0.8,0.92)]
0.8	[(0.0091,0.0123),p(0.7,0.75,0.8)]	[(0.0148,0.0189),p(0.6,0.7,0.8)]
1	[(0.0132,0.0132),p(0.7,0.8,0.9)]	[(0.0710,0.0710),p(0.6,0.75,0.87)]

By utilizing the inverse Sumudu transform, the following is obtained

$$\begin{cases} \underline{x}(t,\alpha) = \underline{x}(0,\alpha)\mathbf{S}^{-1}\left(\frac{1}{B+1}\right) \\ \overline{x}(t,\alpha) = \overline{x}(0,\alpha)\mathbf{S}^{-1}\left(\frac{1}{B+1}\right) \end{cases} \tag{6.154}$$

where

$$\begin{cases} \underline{x}(t,\alpha) = e^{-t}\underline{x}(0,\alpha) \\ \overline{x}(t,\alpha) = e^{-t}\overline{x}(0,\alpha) \end{cases} \tag{6.155}$$

If the initial condition be a symmetric triangular Z-number as $x(0) = [(-a(1-\alpha), a(1-\alpha)), p(0.8, 0.9, 1)]$, then the following cases will hold.
Case 1.

$$\begin{cases} \underline{x}(t,\alpha) = e^{t}(-a(1-\alpha)) \\ \overline{x}(t,\alpha) = e^{t}(a(1-\alpha)) \end{cases}. \tag{6.156}$$

Case 2.

$$\begin{cases} \underline{x}(t,\alpha) = e^{-t}(-a(1-\alpha)) \\ \overline{x}(t,\alpha) = e^{-t}(a(1-\alpha)) \end{cases}. \tag{6.157}$$

Approximation errors based on Z-numbers are shown in Table 6.13. In this table, the results are compared with the average Euler method [185]. ∎

6.10 Summary

Dual fuzzy equations with Z-number coefficients are used to model uncertain nonlinear systems. Then the relation between the solution of the fuzzy equations and the controllers is given. The controllability of the fuzzy system is proposed. Two types of neural network are applied to approximate the solutions of the fuzzy equations. Modified gradient descent algorithms are used to train the neural networks. The new methods are validated by several benchmark examples.

References

1 L.P. Aarts, P. Van der veer, Neural network method for solving partial differential equations, *Neural Processing Letters*, Vol.14, pp.261–271, 2001.

2 K. Abbaoui, Y. Cherruault, Convergence of Adomian's method applied to nonlinear equations, *Math. Comput. Model.* Vol.20, pp.69–73, 1994.

3 S. Abbasbandy, Improving Newton–Raphson method for nonlinear equations by modified Adomian decomposition method, *Appl. Math. Comput.* Vol.145, pp.887–893, 2003.

4 S. Abbasbandy, T. Allahvinloo, Numerical solutions of fuzzy differential equation, *Meth. Comput. Appl.* Vol.7, pp.41–52, 2002.

5 S. Abbasbandy, T. Allahvinloo, Numerical solutions of fuzzy differential equations by Taylor method, *Computational Methods in Applied Mathematics* Vol.2, pp.113–124, 2002.

6 S. Abbasbandy, T. Allahvinloo, Numerical solution of fuzzy differential equation by Runge-Kutta method, *Nonlinear Studies*, Vol.11, pp.117–129, 2004.

7 S. Abbasbandy, B. Asady, Newton's method for solving fuzzy nonlinear equations, *Appl. Math. Comput.* Vol. 159, pp.349–356, 2004.

8 S. Abbasbandy, R. Ezzati, Newton's method for solving a system of fuzzy nonlinear equations, *Appl. Math. Comput.* Vol.175, pp.1189–1199, 2006.

9 S. Abbasbandy, A. Jafarian, Steepest descent method for solving fuzzy nonlinear equations, *Appl. Math. Comput.* Vol.174, pp.669–675, 2006.

10 S. Abbasbandy, M. Otadi, Numerical solution of fuzzy polynomials by fuzzy neural network, *Appl. Math. Comput.* Vol.181, pp.1084–1089, 2006.

11 A. Abdelrazec, D. Pelinovsky, Convergence of the Adomian Decomposition method for Initial-Value problems, *Numerical Methods for Partial Differential Equations*, Vol.27, pp.749–766, 2011.

12 C.D. Acosta, R. Burger, C.E. Mejia, Monotone difference schemes stabilized by discrete mollification for strongly degenerate parabolic equations, *Numer Meth Part Differ Equat.* Vol.28, pp.38–62, 2012.

Modeling and Control of Uncertain Nonlinear Systems with Fuzzy Equations and Z-Number,
First Edition. Wen Yu and Raheleh Jafari.
© 2019 by The Institute of Electrical and Electronics Engineers, Inc. Published 2019 by John Wiley & Sons, Inc.

13 G. Adomian, A review of the Decomposition method in Applied Mathematics, *Journal of Mathematical Analysis and Applicatins*, Vol.135, pp.501–544, 1988.

14 G. Adomian, R. Rach, On the solution of algebraic equations by the decomposition method, *Journal of Mathematical Analysis and Applications*, Vol.105, pp.141–166, 1985.

15 S. Agatonovic-Kustrin, R. Beresford, Basic concepts of artificial neural network (ANN) modeling and its application in pharmaceutical research, *Journal of Pharmaceutical and Biomedical Analysis*. vol.22, pp.717–727, 2000.

16 T. Allahviranloo, M. Otadi, M. Mosleh, Iterative method for fuzzy equations, *Soft Computing*, Vol.12, pp.935–939, 2007.

17 M. Afshar Kermani, F. Saburi, Numerical Method for Fuzzy Partial Differential Equations, *Appl. Math. Sci.* Vol.1, pp.1299–1309, 2007.

18 M.Z. Ahmadi, M.K. Hasan, A new fuzzy version of Euler's method for solving differential equations with fuzzy initial values, *Sains Malaysiana*, Vol.40, pp.651–657, 2011.

19 M.Z. Ahmadi, M.K. Hasan, Incorporating optimisation technique into Euler method for solving differential equations with fuzzy initial values, *Proceeding of the 1st Regional Conference on Applied and Engineering Mathematics*, Malaysia, 2010.

20 M.B. Ahmadi, N.A. Kiani, N. Mikaeilvand, Laplace transform formula on fuzzy nth-order derivative and its application in fuzzy ordinary differential equations, *Soft Comput.* Vol.18, pp.2461–2469, 2014.

21 R.A. Aliev, A.V. Alizadeh, O.H. Huseynov, The arithmetic of discrete Z-numbers, *Inform. Sci.* Vol.290, pp.134–155, 2015.

22 R.A. Aliev, O.H. Huseynov, *Decision theory with imperfect information*, World Scientific, 2014.

23 R.A. Alieva, W. Pedryczb, V. Kreinovich, O.H. Huseynov, The general theory of decisions, *Inform. Sci.* Vol.327, pp.125–148, 2016.

24 T. Allahviranloo, Difference methods for fuzzy partial differential equations, *Computational Methods in Applied Mathematics*, Vol.2, pp.233–242, 2002.

25 T. Allahviranloo, S. Abbasbandy, N. Ahmady, E. Ahmady, Improved predictor-corrector method for solving fuzzy initial value problems, *Inform. Sci.* Vol.179, pp.945–955, 2009.

26 T. Allahviranloo, M.B. Ahmadi, Fuzzy Laplace Transform, *Soft Computing*. Vol.14, pp.235–243, 2010.

27 T. Allahviranloo, N. Ahmadi, E. Ahmadi, Numerical solution of fuzzy differential equations by predictor-corrector method, *Inform. Sci.* Vol.177, pp.1633–1647, 2007.

28 T. Allahviranloo, S. Asari, Numerical solution of fuzzy polynomials by Newton–Raphson method, *J. Appl. Math.* Vol.27, pp.17–24, 2011.

29 T. Allahviranloo, N.A, Kiani, N. Motamedi, Solving fuzzy differential equations by differential transformation method, *Inform. Sci.* Vol.179, pp.956–966, 2009.

30 M. Amirfakhrian, Numerical solution of algebraic fuzzy equations with crisp variable by Gauss–Newton method, *Appl. Math. Model.* Vol.32, pp.1859–1868, 2008.

31 E. Babolian, J. Biazar, Solution of nonlinear equations by modified Adomian decomposition method, *Appl. Math. Comput.* Vol.132, pp.167–172, 2002.

32 E. Babolian, J. Biazar, On the order of convergence of Adomian method, *Appl. Math. Comput.* Vol.130, pp.383–387, 2002.

33 E. Babolian, Sh. Javadi, Restarted Adomian method for algebraic equations, *Appl. Math. Comput.* Vol.146, pp.533–541, 2003.

34 M.R. Balooch Shahryari, S. Salahshour, Improved predictor-corrector method for solving fuzzy differential equations under generalized differentiability, *Journal of Fuzzy Set Valued Analysis*, Vol.2012, pp.1–16, 2012.

35 A. Ban, B. Bede, Properties of the cross product of fuzzy numbers, *Journal of Fuzzy Mathematics*, Vol.14, pp.513–531, 2006.

36 M. Barkhordari Ahmadi, N.A. Kiani, Solving fuzzy partial differential equation by differential transformation method, *Journal of Appl. Math.* Vol.7, pp.1–16, 2011.

37 B. Bede, Note on numerical solutions of fuzzy differential equations by predictor corrector method, *Inform. Sci.* Vol.178, pp.1917–1922, 2008.

38 B. Bede, S.G. Gal, Generalizations of differentiablity of fuzzy number valued function with application to fuzzy differential equations, *Fuzzy Sets Syst.* Vol.151, pp.581–599, 2005.

39 B. Bede, J. Imre, C. Rudas, L. Attila, First order linear fuzzy differential equations under generalized differentiability, *Inform. Sci.* Vol.177, pp.3627–3635, 2007.

40 B. Bede, I.J. Rudas, A.L. Bencsik, First order linear fuzzy differential equations under generalized differentiability, *Inform. Sci.* Vol.177, pp.1648–1662, 2006.

41 B. Bede, L. Stefanini, Generalized differentiability of fuzzy-valued functions, *Fuzzy Sets Syst.* Vol.230, pp.119–141, 2013.

42 F.P. Beer, E.R. Johnston, *Mechanics of materials*, 2nd Edition, mcgraw-Hill. 1992.

43 S.S. Behzadi, T. Allahviranloo, Solving fuzzy differential equations by using picard method, *Iranian Journal of Fuzzy Systems*. Vol.13, pp.71–81, 2016.

44 F.B.M. Belgacem, A.A. Karaballi, Sumudu transform fundamental properties investigations and applications, *J. Appl. Math. Stoch. Anal.* Vol.2006, doi. 10.1155/JAMSA/2006/91083, 2006.

45 A. Bonarini, G. Bontempi, A qualitative simulation approach for fuzzy dynamical models, *ACM Transaction Modelling and Computer Simulation*, Vol.4, pp.285–313, 1994.

46 O. Brudaru, F. Leon, O. Buzatu, Genetic algorithm for solving fuzzy equations, *8th International Symposium on Automatic Control and Computer Science, Iasi, ISBN*, 2004.

47 J.J. Buckley, E. Eslami, Neural net solutions to fuzzy problems: The quadratic equation, *Fuzzy Sets Syst.* Vol.86, pp.289–298, 1997.

48 J.J. Buckley, E. Eslami, T. Feuring, *Fuzzy mathematics in economics and engineering, Studies in fuzziness and soft computing*, Phisica-Verlag, 2002.

49 J.J. Buckley, E. Eslami, Y. Hayashi, Solving fuzzy equations using neural nets, *Fuzzy Sets Syst.* Vol.86, pp.271–278, 1997.

50 J.J. Buckley, T. Feuring, Y. Hayashi, Solving fuzzy equations using evolutionary algorithms and neural nets, *Soft comput.* Vol.6, pp.116–123, 2002.

51 J.J. Buckley, Y. Hayashi, Can fuzzy neural nets approximate continuous fuzzy functions, *Fuzzy Sets Syst.* Vol.61, pp.43–51, 1994.

52 J.J. Buckley, Y. Qu, Solving linear and quadratic fuzzy equations, *Fuzzy Sets Syst.* Vol.35, pp.43–59, 1990.

53 R.L. Burden, J.D. Faires, *Numerical Analysis*, seventh ed., PWS-Kent, Boston, 2001.

54 R.L. Burden, J.D. Faires, *Numerical Analysis*, Thomson, Pacific Grove, Calif, USA, 8th edition, 2005.

55 H.B. Cary, S.C. Helzer, *Modern Welding Technology*. Upper Saddle River, New Jersey: Pearson Education. 2005.

56 Y. Chalco-Cano, H. Roman-Flores, On new solutions of fuzzy differential equations, *Chaos Solitons Fractals*. Vol.38, pp.112–119, 2006.

57 J.-C. Chang, H. Chen, S.-M. Shyu, W.-C. Lian, Fixed-point theorems in fuzzy real line, *Comput. Math. Appl.* Vol.47, pp.845–851, 2004.

58 S.M. Chen, Fuzzy system reliability analysis using fuzzy number arithmetic operations, *Fuzzy Sets Syst.* Vol.64, pp.31–38, 1994.

59 Y. Cherruault, Convergence of Adomian's method, *Kybernetes*, Vol.9, pp.31–38, 1988.

60 S. Curtis, S. Ghosh, A variable selection approach to monotonic regression with Bernstein polynomials, *J. Appl. Stat.* Vol.38, pp.961–976, 2011.

61 G. Cybenko, Approximation by Superposition of Sigmoidal Activation Function, *Math. Control, Sig Syst*, Vol.2, 303–314, 1989.

62 P.J. Davis, *Interpolation and approximation*, Dover Publications, Inc., New York, 1975.

63 M. Dehghan, On the solution of an initial-boundary value problem that combines neumann and integral condition for the wave equation, *Numer. Methods. Partial. Differ. Equ.* Vol.21, pp.24–40, 2005.

64 M. Delgado, M.A. Vila, W. Voxman, On a canonical representation of fuzzy Numbers, *Fuzzy Sets Syst.* Vol.93, pp.125–135, 1998.

65 P. Diamond, Fuzzy least squares, *Inform. Sci.* Vol.46, pp.141–157, 1988.

66 M.W.M.G. Dissanayake, N. Phan-Thien, Neural-network based approximations for solving partial differential equations, *Communications in Numerical Methods in Engineering.* Vol.10, pp.195–201, 2000.

67 S. Effati, M. Pakdaman, Artificial neural network approach for solving fuzzy differential equations, *Inform. Sci.* Vol.180, pp.1434–1457, 2010.

68 A. Farajzadeh, A. Hossein Pour, M. Amini, An explicit method for solving fuzzy partial differential equation, *International Mathematical Forum*, Vol.5, pp.1025–1036, 2010.

69 G. Feng, A survey on analysis and design of model-based fuzzy control systems, *IEEE Trans. Fuzzy Syst.* Vol.14, pp.676–697, 2006.

70 M. Friedman, M. Ming, A. Kandel, Fuzzy linear systems, *Fuzzy Sets Syst.* Vol.96, pp.201–209, 1998.

71 M. Friedman, M. Ming, A. Kandel, Duality in fuzzy linear systems, *Fuzzy Sets Syst.* Vol.109, pp.55–58, 2000.

72 L.A. Gardashova, Application of operational approaches to solving decision making problem using Z-Numbers, *Journal of Applied Mathematics.* Vol.5, pp.1323–1334, 2014.

73 B. Ghazanfari, A. Shakerami, Numerical solutions of fuzzy differential equations by extended Runge–Kutta-like formulae of order 4, *Fuzzy Sets Syst.* Vol.189, pp.74–91, 2011.

74 J.S. Gibson, An analysis of optimal modal regulation: convergence and stability, *SIAM J. Control. Optim.* Vol.19, pp.686–707, 1981.

75 S. Guo, L. Mei, Y. Zhou, The compound $\frac{G'}{G}$-expansion method and double non-traveling wave solutions of (2+1)-dimensional nonlinear partial differential equations, *COMPUT. MATH. APPL.* Vol.69, pp.804–816, 2015.

76 E. Hajilou, M. Paripour, H. Heidari, Application of differential transform method to solve hybrid fuzzy differential equations, *International Journal of Mathematical Modelling and Computations*, Vol.5, pp.203–217, 2015.

77 Y.M. Ham, C. Chun, S.G. Lee, Some higher-order modifications of Newton's method for solving nonlinear equations, *J. Comput. Appl. Math.* Vol. 222, pp.477–486, 2008.

78 M. Hazewinkel, *Oscillator harmonic*, Springer, ISBN, 2001.

79 S. He, K. Reif, R. Unbehauen, Multilayer neural networks for solving a class of partial differential equations, *Neural Networks.* Vol.13 pp.385–396, 2000.

80 H. Holden, K.H. Karlsen, K.A. Lie, Operator splitting methods for degenerate convection-diffusion equations II: numerical examples with emphasis on reservoir simulation and sedamentation, *Comput. Geosci.* Vol.4, pp.287–322, 2000.

81 J.P. Holman, *Heat transfer*, 8th Ed, McGraw-Hill. New York. 1997.

82 T.J.R. Hughes, *The finite element method*, Prentice Hall, New Jersey, 1987.

83 E. Hüllermeier, An approach to modeling and simulation of uncertain dynamical systems, *Internat. J. Uncertainty Fuzzyness Knowledge-Based Syst.* Vol.5, pp.117–137, 1997.

84 E.A. Hussian, M.H. Suhhiem, Numerical solution of fuzzy differential equations based on Taylor series by using fuzzy neural networks, *advance in mathematics*, Vol.11, pp.4080–4092, 2015.

85 E.A. Hussian, M.H. Suhhiem, Numerical solution of fuzzy differential equations by using modified fuzzy neural network, *International Journal of Mathematical Archive*, Vol.6, pp.84–94, 2015.

86 E.A. Hussian, M.H. Suhhiem, Numerical solution of fuzzy partial differential Equations by using modified fuzzy neural networks, *British Journal of Mathematics and Computer Science*, Vol.12, pp.1–20, 2016.

87 J.M. Hyman, M. Shashkov, The approximation of boundary conditions for mimetic finite difference methods, *Computers Math. Applic.* Vol.36, pp.79–99, 1998.

88 R. Jafari, S. Razvarz, Solution of Fuzzy Differential Equations using Fuzzy Sumudu Transforms, *IEEE International Conference on Innovations in Intelligent Systems and Applications*, pp.84–89, 2017.

89 R. Jafari, S. Razvarz, A. Gegov, A new computational method for solving fully fuzzy nonlinear systems, *Computational Collective Intelligence: 10th International Conference, ICCCI 2018, Bristol, UK, September 5-7, 2018, Proceedings, Part I, Lecture Notes in Computer Science, Springer*, Vol.11055, pp.503–512, 2018.

90 R. Jafari, S. Razvarz, A. Gegov, S. Paul, Fuzzy modeling for uncertain nonlinear systems using fuzzy equations and Z-numbers. *Advances in Computational Intelligence Systems: Contributions Presented at the 18th UK Workshop on Computational Intelligence, September 5-7, 2018, Nottingham, UK. Advances in Intelligent Systems and Computing, Springer*, Vol.840, pp.66–107, 2018.

91 R. Jafari, W. Yu, Artificial neural network approach for solving strongly degenerate parabolic and burgers-fisher equations, *12th International Conference on Electrical Engineering, Computing Science and Automatic Control*, doi:10.1109/ICEEE.2015.7357914, 2015.

92 R. Jafari, W. Yu, Uncertainty nonlinear systems modeling with fuzzy equations, In *Proceedings of the 16th IEEE International Conference on Information Reuse and Integration (IRI '15)*, pp.182–188, San Francisco, Calif, USA, August 2015.

93 R. Jafari, W. Yu, Uncertainty Nonlinear Systems Control with Fuzzy Equations, *IEEE International Conference on Systems, Man, and Cybernetics*, pp.2885–2890, 2015.

94 R. Jafari, W. Yu, Uncertainty Nonlinear Systems Modelling with Fuzzy Equations, *IEEE International Conference on Information Reuse and Integration*, pp.182–188, 2015.

95 R. Jafari, W. Yu, Fuzzy Control for Uncertainty Nonlinear Systems with Dual Fuzzy Equations, *Journal of Intelligent and Fuzzy Systems*. Vol.29, pp.1229–1240, 2015.

96 R. Jafari, W. Yu, Uncertain nonlinear system control with fuzzy differential equations and Z-numbers, *18th IEEE International Conference on Industrial Technology, Canada*, pp.890 -895, doi:10.1109/ICIT.2017.7915477, 2017.

97 R. Jafari, W. Yu, Fuzzy Modeling for Uncertainty Nonlinear Systems with Fuzzy Equations, *Mathematical problems in Engineering*. Vol.2017, https://doi .org/10.1155/2017/8594738, 2017.

98 R. Jafari, W. Yu, Fuzzy Differential Equation for Nonlinear System Modeling with Bernstein Neural Networks, *IEEE Access*. doi:10.1109/ACCESS.2017.2647920, 2017.

99 R. Jafari, W. Yu, X. Li, Solving Fuzzy Differential Equation with Bernstein Neural Networks, *IEEE International Conference on Systems, Man, and Cybernetics, Budapest, Hungary*, pp.1245–1250, 2016.

100 R. Jafari, W. Yu, X. Li, Numerical solution of fuzzy equations with Z-numbers using neural networks, *Intelligent automation and Soft Computing*, pp.1–7, 2017.

101 R. Jafari, W. Yu, X. Li, S. Razvarz, Numerical Solution of Fuzzy Differential Equations with Z-numbers Using Bernstein Neural Networks, *International Journal of Computational Intelligence Systems*, Vol.10, pp.1226–1237, 2017.

102 A. Jafarian, R. Jafari, Approximate solutions of dual fuzzy polynomials by feed-back neural networks, *J. Soft Comput. Appl.* 2012, doi:10.5899/2012/jsca-00005.

103 A. Jafarian, R. Jafari, A. Khalili, D. Baleanud, Solving fully fuzzy polynomials using feed-back neural networks, *International Journal of Computer Mathematics*, Vol.92, pp.742–755, 2015.

104 A. Jafarian, R. Jafari, M. Mohamed Al Qurashi. D. Baleanud, A novel computational approach to approximate fuzzy interpolation polynomials, *Springer Plus*. Vol.5, pp.14–28, 2016.

105 T. Jayakumar, D. Maheskumar, K. Kanagarajan, Numerical solution of fuzzy differential equations by Runge Kutta method of order five, *Appl. Math. Sci.* Vol.6, pp.2989–3002, 2012.

106 T. Jayakumar, K. Kanagarajan, S. Indrakumar, Numerical solution of Nth-order fuzzy differential equation by Runge-Kutta Nystrom method, *International Journal of Mathematical Engineering and Science*, Vol.1, pp.2277–6982, 2012.

107 U. Kadak, F. Basar, On some sets of fuzzy-valued sequences with the level sets, *Contemporary Analysis and Applied M athematics*, Vol.1, pp.70–90, 2013.

108 M. Kajani, B. Asady, A. Vencheh, An iterative method for solving dual fuzzy nonlinear equations. *Appl. Math. Comput.* Vol.167, pp.316–323, 2005.

109 K. Kanagarajan, M. Sambath, Runge-Kutta Nystrom method of order three for solving fuzzy differential equations, *Computational Methods in Applied Mathematics*, Vol.10, pp.195–203, 2010.

110 B. Kang, D. Wei, Y. Li, Y. Deng, A method of converting Z-number to classical fuzzy number, *Journal of Information and Computational Science*. Vol.9, pp.703–709, 2012.

111 B. Kang, D. Wei, Y. Li, Y. Deng, Decision making using Z–Numbers under uncertain environemnt, *Journal of Computational Information Systems*. Vol.8, pp.2807–2814, 2012.

112 K. Kemati, M. Matinfar, An implicit method for fuzzy parabolic partial differential equations, *J. Nonlinear Sci. Appl.* Vol.1, pp.61–71, 2008.

113 A. Khastan, K. Ivaz, Numerical solution of fuzzy differential equations by Nyström method, *Chaos, Solitons. Fractals.* Vol. 41, pp.859–868, 2009.

114 H. Kim, R. Sakthivel, Numerical solution of hybrid fuzzy differential equations using improved predictor-corrector method, *Commun Nonlinear Sci Numer Simulat*, Vol.17, pp.3788–3794, 2012.

115 A. Kröner, K. Kunisch, A minimum effort optimal control problem for the wave equation, *Comput. Optim. Appl.* Vol.57, pp.241–270, 2014.

116 I.E. Lagaris, A. Likas, D.I. Fotiadis, Artificial neural networks for solving ordinary and partial differential equations, *IEEE Transactions on Neural Networks*. Vol.9, pp.987–1000, 1998.

117 V. Laksmikantham, R.N. Mohapatra, *Theory of Fuzzy Differential Equations and Inclusions*, Taylor and Francis, NewYork, 2003.

118 H. Lee, I.S. Kang, Neural algorithms for solving differential equations, *Journal of Computational Physics*. Vol.91, pp.110–131, 1990.

119 R.J. Leveque, Finite difference methods for differential equations, *University of Washington*, 2005.

120 F.T Lin, simulating fuzzy numbers for solving fuzzy equations with constraints using genetic algorithm, *International Journal of Innovative Computing, Information and Control* Vol.6, pp.239–253, 2010.

121 M.A. Ma, M. Friedman, A. Kandel, Numerical solutions of fuzzy differential equations, *Fuzzy Sets Syst.* Vol.105, pp.133–138, 1999.

122 E.H. Mamdani, Application of fuzzy algorithms for control of simple dynamic plant, *IEE Proceedings-Control Theory and Applications.* Vol.121, pp.1585–1588, 1976.

123 P. Mansouri, B. Asady, N. Gupta, An approximation algorithm for fuzzy polynomial interpolation with Artificial Bee Colony algorithm, *Appl. Soft. Comput.* Vol.13, pp.1997–2002, 2013.

124 M.H. Mashinchi, M.R. Mashinchi, S. Mariyam, H.J. Shamsuddin, A genetic algorithm approach for solving fuzzy linear and quadratic equations, *World Academy of Science, Engineering and Technology*, Vol.1, No:4, 2007.

125 N.E. Mastorakis, Solving non-linear equations via genetic algorithms, *6th WSEAS Int. Conf. on EVOLUTIONARY COMPUTING, Lisbon, Portugal*, pp.24–28, 2005.

126 H.N. Mhaskar, D.V. Pai, *Fundamentals of Approximation Theory*, CRC Press, Boca Raton. FL, 2000.

127 Z. Michalewicz, *Genetic algorithms+data structures=evolution program*, 2nd ed., Artificial Intelligence, Springer Verlag, Berlin, 1994.

128 N. Mikaeilvand, S. Khakrangin, Solving fuzzy partial differential equations by fuzzy two-dimensional differential transform method, *Neural. Comput. Appl.* Vol.21, pp.307–312, 2012.

129 C. Montelora, C. Saloma, Solving the nonlinear schrodinger equation with an unsupervised neural network: Estimation of error in solution, *Opt. Commun.* Vol.222, pp.331–339, 2003.

130 M.K. Moraveji, S. Razvarz, Experimental investigation of aluminum oxide nanofluid on heat pipe thermal performance, *Int. Commun. Heat. Mass. Transf.* Vol.39, pp.1444–1448, 2012.

131 M. Mosleh, Solution of dual fuzzy polynomial equations by modified Adomian decomposition method, *Fuzzy Inf. Eng.* Vol.1, pp.45–56, 2013.

132 M. Mosleh, Evaluation of fully fuzzy matrix equations by fuzzy neural network, *Appl. Math. Model.* Vol.37, pp.6364–6376, 2013.

133 M. Mosleh, M. Otadi, A new approach to the numerical solution of dual fully fuzzy polynomial equations, *Int. J. Industrial Mathematic.* Vol.2, pp.129–142, 2010.

134 M. Mosleh, M. Otadi, S. Abbasbandy, Fuzzy polynomial regression with fuzzy neural networks, *Appl. Math. Model.* Vol.35, pp.5400–5412, 2011.

135 M. Mosleh, M. Otadi, Simulation and evaluation of fuzzy differential equations by fuzzy neural network, *Appl. Soft. Comput.* Vol.12, pp.2817–2827, 2012.

136 M. Mosleh, M. Otadi, Solving the second order fuzzy differential equations by fuzzy neural network, *Journal of Mathematical Extension* Vol.8, pp.11–27, 2014.

137 E. Nasrabadi, Fuzzy linear regression models analysis: mathematical-programming methods, *M.Sc. Thesis, Department of Industrial Engineering, Sharif University of Technology, Tehran*, 2003.

138 M.M. Nasrabadi, E. Nasrabadi, A.R. Nasrabadi, Fuzzy linear regression analysis: a multi-objective programming approach, *Appl. Math. Comput.* Vol. 163, pp.245–251, 2005.

139 K. Nemati, M. Matinfar, An implicit method for fuzzy parabolic partial differential equations, *J. Nonlinear Sci. Appl.* Vol.1, pp.61–71, 2008.

140 I. Newton, Methodus fluxionum et serierum infinitarum, 1664–1671.

141 H.T. Nguyen, A note on the extension principle for fuzzy sets, *J. Math. Anal. Appl.* Vol.64, pp.369–380, 1978.

142 J.J. Nieto, A. Khastan, K. Ivaz, Numerical solution of fuzzy differential equations under generalized differentiability, *Nonlinear Anal. Hybrid Syst.* Vol.3, pp.700–707, 2009.

143 A. Seyed Alavi Nobar, Convergence and stability properties Euler method for solving fuzzy Stochastic differential equations, *Journal of Fuzzy Set Valued Analysis*, Vol.2015, pp.194–207, 2015.

144 A. Noorani, J. Kavikumar, M. Mustafa, and S. Nor, Solving dual fuzzy polynomial equation by ranking method, *Far East J. Math. Sci.* Vol.15, pp.151–163, 2011.

145 R. Nurhakimah Ab, A. Lazim, Solving Fuzzy Polynomial Equation and The Dual Fuzzy Polynomial Equation Using The Ranking Method, *The 3th International conference on Mathematics Scinces* pp. 798–804, 2014.

146 R. Nurhakimah Ab, A. Lazim, An Interval Type-2 Dual Fuzzy Polynomial Equations and Ranking Method of Fuzzy Numbers, *International Journal of Mathematical, Computational, Physical, Electrical and Computer Engineering*, Vol.8, 2014.

147 R. Nurhakimah Ab, A. Lazim, Ab.G. Ahmad Termimi, A. Noorani, Solutions of Interval Type-2 Fuzzy Polynomials Using A New Ranking Method, *AIP Conference*, Vol.1682, 2015.

148 M. Otadi, Fully fuzzy polynomial regression with fuzzy neural networks, *Neurocomputing*, Vol.142, pp.486–493, 2014.

149 M. Otadi, M. Mosleh, Solution of fuzzy polynomial equations by modified Adomian decomposition method, *Soft. Comput.* Vol.15, pp.187–192, 2011.

150 M. Otadi, M. Mosleh, S. Abbasbandy, Numerical solution of fully fuzzy linear systems by fuzzy neural network, *Soft Comput.* Vol.15, pp.1513–1522, 2011.

151 S.C. Palligkinis, G. Papageorgiou, I.T. Famelis, Runge-Kutta methods for fuzzy differential equations, *Applied Mathematics and Computation* Vol.209, pp.97–105, 2009.

152 M. Paripour, E. Hajilou, H. Heidari, Application of Adomian decomposition method to solve hybrid fuzzy differential equations, *Journal of Taibah University for Science*, http://dx.doi.org/10.1016/j.jtusci .2014.06.002, 2014.

153 S. Pederson, M. Sambandham, Numerical solution to hybrid fuzzy systems. *Math. Comput. Model.* Vol.45, pp.1133–1144, 2007.

154 S. Pederson, M. Sambandham, The Runge-Kutta method forhybrid fuzzy differential equation, *Nonlinear Anal. Hybrid Syst.* Vol.2, pp.626–634, 2008.

155 G. Peters, Fuzzy linear regression with fuzzy intervals, *Fuzzy Sets Syst.* Vol.63, pp.45–55, 1994.

156 U.M. Pirzada, D.C. Vakaskar, Solution of fuzzy heat equations using Adomian Decomposition method, *Int. J. Adv. Appl. Math. and Mech.* Vol.3, pp.87–91, 2015.

157 R.H. Pletcher, J.C. Tannehill, D. Anderson, *Computational Fluid Mechanics and Heat Transfer*, Taylor and Francis, 1997.

158 H. Pohlheim, Documentation for genetic and evolutionary algorithm toolbox for use with matlab (GEATbx), version 1.92, http://www.geatbx.com, 1999.

159 H. Radstrom, An embedding theorem for spaces of convex sets, *Proc. Amer. Math. Soc.* Vol.3, pp.165–169, 1952.

160 D. Ralescu, Toward a general theory of fuzzy variables, *Journal of Mathematical Analysis and Applications*, Vol.86, pp.176–193, 1982.

161 A. Ramli, M.L. Abdullah, M. Mamat, Broyden's method for solving fuzzy nonlinear equations, *Advances in Fuzzy Systems*, Vol.2010, Article ID 763270, 6 pages, doi:10.1155/2010/763270.

162 S. Razvarz, R. Jafari, ICA and ANN Modeling for Photocatalytic Removal of Pollution in Wastewater, *Mathematical and Computational Applications*, Vol.22, pp.38–48, 2017.

163 S. Razvarz, R. Jafari, Intelligent Techniques for Photocatalytic Removal of Pollution in Wastewater,*Journal of Electrical Engineering*, Vol.5, pp.321–328, 2017. doi: 10.17265/2328-2223/2017.06.004.

164 S. Razvarz, R. Jafari, A. Gegov, W. Yu, S. Paul, *Neural network approach to solving fully fuzzy nonlinear systems, Fuzzy modeling and control Methods Application and Research*, Nova science publisher, Inc, NewYork. ISBN: 978-1-53613-415-5, pp.45–68, 2018.

165 S. Razvarz, R. Jafari, O.C. Granmo, A. Gegov, Solution of dual fuzzy equations using a new iterative method, In *Proceedings of the 10th Asian Conference on Intelligent Information and Database Systems, Lecture Notes in Artificial Intelligence (subseries of LNCS)*. Springer,doi: 10.1007/978-3-319-75420-8-23, pp.245–255, 2018

166 S. Razvarz, R. Jafari, W. Yu, Numerical Solution of Fuzzy Differential Equations with Z-numbers using Fuzzy Sumudu Transforms, *Advances in Science, Technology and Engineering Systems Journal (ASTESJ)*, Vol.3, pp.66–75, doi: 10.25046/aj030108, 2018.

167 S. Razvarz, R. Jafari, W. Yu, A. Khalili, PSO and NN Modeling for Photocatalytic Removal of Pollution in Wastewater, *14th International Conference on Electrical Engineering, Computing Science and Automatic Control (CCE) Electrical Engineering*, pp.1–6, 2017.

168 S. Razvarz, M. Tahmasbi, Fuzzy equations and Z-numbers for nonlinear systems control, *Procedia Computer Science*, Vol.120, pp.923–930, 2017.

169 H. Roman-Flores, M. Rojas-Medar, Embedding of level-continuous fuzzy sets on Banach spaces, *Inform. Sci.* Vol.144, pp.227–247, 2002.

170 H. Rouhparvar, Solving fuzzy polynomial equation by ranking method, First Joint Congress on Fuzzy and Intelligent Systems, *Ferdowsi University of Mashhad, Iran*, 2007.

171 Xu. Ruoning, S-curve regression model in fuzzy environment, *Fuzzy Sets Syst.* Vol.90, pp.317–326, 1997.

172 M. Sakawa, H. Yano, Multi objective fuzzy linear regression analysis for fuzzy input–output data, *Fuzzy Sets Syst.* Vol.47, pp.173–181, 1992.

173 D. Savic and W. Pedryzc, Evaluation of fuzzy linear regression models, *Fuzzy Sets Syst.* Vol.39, pp.51–63, 1991.

174 P. Sevastjanov, L. Dymova, A new method for solving interval and fuzzy equations: Linear case, *Inform. Sci.* Vol.179, pp.925–937, 2009.

175 O. Solaymani Fard, T. AliAbdoli Bidgoli, The Nyström method for hybrid fuzzy differential equation IVPs, *Journal of King Saud University-Science*, Vol.23, pp.371–379, 2011.

176 M.W. Spong, M. Vidyasagar, *Robot Dynamics and Control*, John Wiley and Sons, 1989.

177 V.L. Streeter, E.B. Wylie, E. Benjamin, *Fluid mechanics*, 4th Ed, Mc. Graw-Hill Book Company. 1999.

178 P.V. Subrahmanyam, S.K. Sudarsanam, On some fuzzy functional equations, *Fuzzy Sets Syst.* Vol.64, pp.333–338, 1994.

179 N. Sukavanam, V. Panwar, Computation of boundary control of controlled heat equation using artificial neural networks, *Int. Commun. Heat Mass Transfer.* Vol.30, pp.1137–1146, 2003.

180 J.A.K. Suykens, J. De Brabanter, L. Lukas, J. Vandewalle, Weighted least squares support vector machines: robustness and sparse approximation, *Neurocomputing*, Vol.48 pp.85–105, 2002.

181 A. Tahavvor, M. Yaghoubi, Analysis of natural convection from a column of cold horizontal cylinders using artificial neural network, *Appl. Math. Model.* Vol.36, pp.3176–3188, 2012.

182 T. Takagi, M. Sugeno, Fuzzy identification of systems and its applications to modeling and control, *IEEE Trans. Syst., Man, Cybern.* Vol.15, pp.116–132, 1985.

183 H. Tanaka, I. Hayashi, J. Watada, Possibilistic linear regression analysis for fuzzy data, *Eur. J. Oper. Res.* Vol.40, pp.389–396, 1989.

184 H. Tanaka, S. Uegima, K. Asai, Linear regression analysis with fuzzy model, *IEEE Trans. Syst. ManCybern.* Vol.12, pp.903–907, 1982.

185 S. Tapaswini, S. Chakraverty, Euler-based new solution method for fuzzy initial value problems, *Int. J. Artificial. Intell. Soft. Comput.* Vol.4, pp.58–79, 2014.

186 S. Tapaswini, S. Chakraverty, A new approach to fuzzy initial value problem by improved Euler method, *Fuzzy Inf. Eng.* Vol.3, pp.293–312, 2012.

187 M. Tatari, B. Sepehrian, M. Alibakhshi, New implementation of radial basis functions for solving Burgers-Fisher equation, *Numer. meth. part. Differ. Equat.* Vol.28, pp.248–262, 2010.

188 B.H. Tongue, *Principles of Vibration*, Oxford University Press, 2001.

189 A.R. Vahidi, B. Jalalvand, Improving the Accuracy of the Adomian Decomposition Method for Solving Nonlinear Equations, *Appl. Math. Sci.* Vol.6, pp.487–497, 2012.

190 M. Wagenknecht, R.Hampel, V.Schneider, Computational aspects of fuzzy arithmetics based on archimedean t-norms, *Fuzzy Sets Syst.* Vol.123, pp.49–62, 2001.

191 H.F. Wang, R.C. Tsaur, Resolution of fuzzy regression model, *Europ. J. Oper. Res.* Vol.126, pp.637–650, 2000.

192 M. Waziri, Z. Majid, A new approach for solving dual fuzzy nonlinear equations using Broyden's and Newton's methods, *Advances in Fuzzy Systems*, Vol.2012, Article 682087, 5 pages, 2012.

193 L.B. Yang, Y.Y. Gao, Fuzzy Mathematics-Theory and its Application, *published by South China University of Technology, Guangzhou China*, 1993.

194 H.S. Yazdi, R. Pourreza, Unsupervised adaptive neural-fuzzy inference system for solving differential equations, *Appl. Soft. Comput.* Vol.10, pp.267–275, 2010.

195 W. Yu, X. Li, Fuzzy identification using fuzzy neural networks with stable learning algorithms, *IEEE Transactions on Fuzzy Systems*, Vol.12, pp.411–420, 2004.

196 L.A. Zadeh, The concept of a liguistic variable and its application to approximate reasoning, *Inform. Sci.* Vol.8, pp.199–249, 1975.

197 L.A. Zadeh, Toward a generalized theory of uncertainty (GTU) an outline, *Inform. Sci.* Vol.172, pp.1–40, 2005.

198 L.A. Zadeh, Generalized theory of uncertainty (GTU)-principal concepts and ideas, *Computational Statistics and Data Analysis*. Vol.51, pp.15–46, 2006.

199 L.A. Zadeh, A note on Z-numbers, *Inform. Sci.* Vol.181, pp.2923–2932, 2011.

Index

Modeling and Control of Uncertain Nonlinear Systems with Fuzzy Equations and Z-Number,
First Edition. Wen Yu and Raheleh Jafari.
© 2019 by The Institute of Electrical and Electronics Engineers, Inc. Published 2019 by John Wiley & Sons, Inc.

IEEE PRESS SERIES ON SYSTEMS SCIENCE AND ENGINEERING

Editor:
MengChu Zhou, *New Jersey Institute of Technology and Tongji University*

Co-Editors:
Han-Xiong Li, *City University of Hong-Kong*
Margot Weijnen, *Delft University of Technology*

The focus of this series is to introduce the advances in theory and applications of systems science and engineering to industrial practitioners, researchers, and students. This series seeks to foster system-of-systems multidisciplinary theory and tools to satisfy the needs of the industrial and academic areas to model, analyze, design, optimize and operate increasingly complex man-made systems ranging from control systems, computer systems, discrete event systems, information systems, networked systems, production systems, robotic systems, service systems, and transportation systems to Internet, sensor networks, smart grid, social network, sustainable infrastructure, and systems biology.

1. *Reinforcement and Systemic Machine Learning for Decision Making*
   Parag Kulkarni

2. *Remote Sensing and Actuation Using Unmanned Vehicles*
   Haiyang Chao and YangQuan Chen

3. *Hybrid Control and Motion Planning of Dynamical Legged Locomotion*
   Nasser Sadati, Guy A. Dumont, Kaveh Akbari Hamed, and William A. Gruver

4. *Modern Machine Learning: Techniques and Their Applications in Cartoon Animation Research*
   Jun Yu and Dachen Tao

5. *Design of Business and Scientific Workflows: A Web Service-Oriented Approach*
   Wei Tan and MengChu Zhou

6. *Operator-based Nonlinear Control Systems: Design and Applications*
   Mingcong Deng

7. *System Design and Control Integration for Advanced Manufacturing*
   Han-Xiong Li and XinJiang Lu

8. *Sustainable Solid Waste Management: A Systems Engineering Approach*
   Ni-Bin Chang and Ana Pires